W9-BXT-203

INVENTING MODERN AMERICA

6
22
10
26
14
28
13
12
S
2
20
F
43
44
34
2
46

INVENTING MODERN AMERICA

From the Microwave to the Mouse

Text by **David E. Brown**
Foreword by **Lester C. Thurow**
Introductions by **James Burke**

A Publication of the Lemelson-MIT Program
for Invention and Innovation

The MIT Press
Cambridge, Massachusetts
London, England

This book was set in Scala Sans and Officina Sans by The MIT Press and printed and bound in the United States of America.

Library of Congress Cataloging-in-Publication Data

Brown, David E.

Inventing Modern America: from the microwave to the mouse / text by David E. Brown ; foreword by Lester C. Thurow ; introductions by James Burke.

p. cm.

"A publication of the Lemelson-MIT program for invention and innovation." Includes bibliographical references and index.

ISBN 0-262-02508-6 (hc.: alk. paper)

1. Inventions—United States—History—20th century. 2. Inventors—United States—Biography. I. Title.

T20 .B76 2002

609.73'09'04—dc21 2001044768

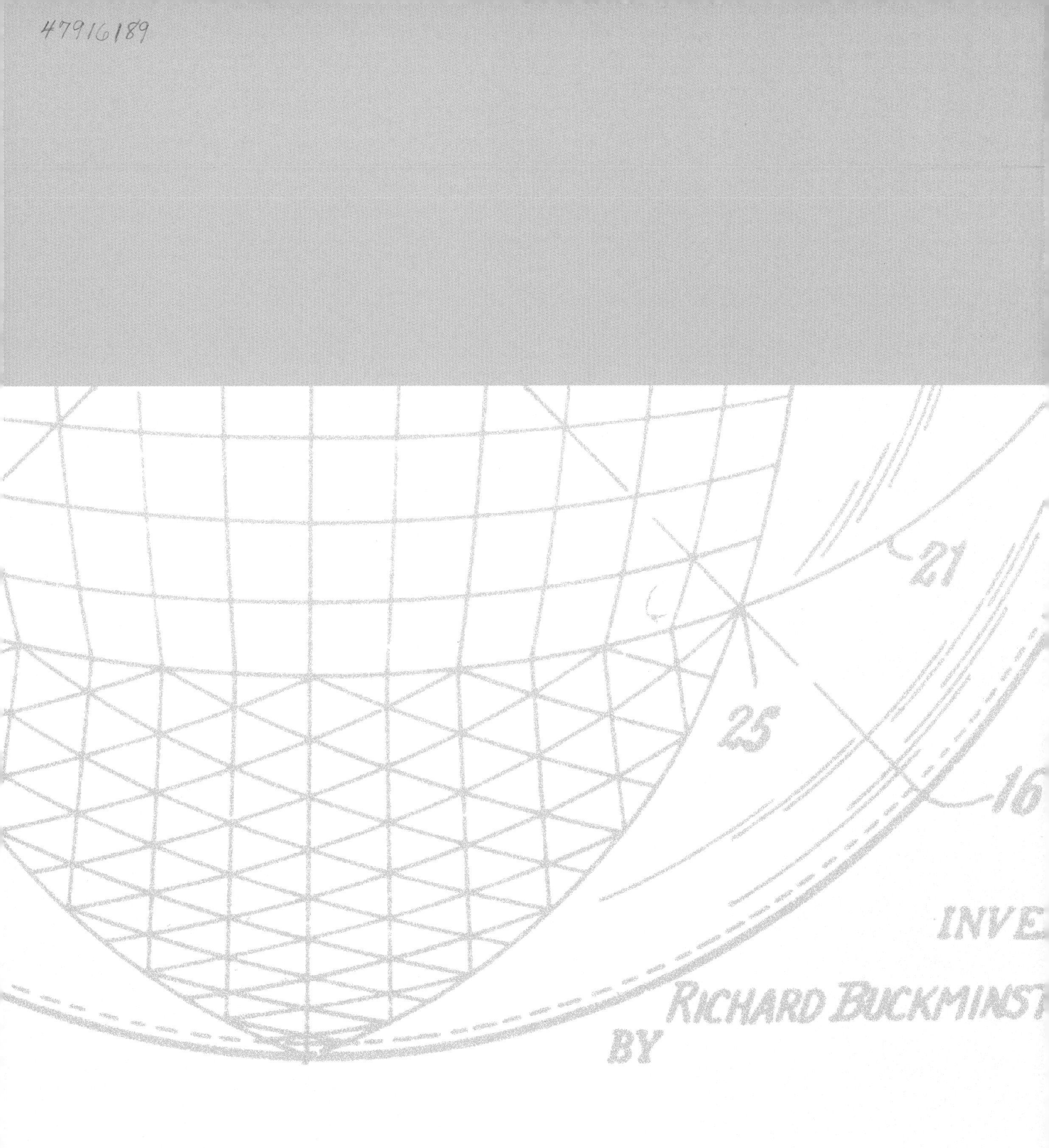
21
25
16
INVE
BY RICHARD BUCKMINST

CONTENTS

TRANSPORTATION

ENERGY AND ENVIRONMENT

COMPUTING AND TELECOMMUNICATIONS

FOREWORD

Lester C. Thurow

Our advance from caves to cyberspace is a history of technical progress. The invention of tools allowed early humans to compete with animals far stronger and faster than they. With humankind's first great technological leap, the domestication of animals and the development of farming, a nomadic hunter-gatherer way of life came to an end. Settled permanent villages emerged. Eventually agriculture had advanced to a point where it could generate food surpluses large enough to feed substantial populations and those villages became cities.

Once everyone did not have to work full time as a farmer, some individuals could devote their time to learning. With the invention of writing and reading, humans took a second great technological leap forward. Knowledge could be saved from one generation to the next. It did not depend upon fallible human memories. It did not depend upon direct person-to-person contact. Knowledge could be accumulated. Information could be sent from one place to another. The dissemination of knowledge was speeded up. Henceforth no one would have to reinvent the wheel. They could read about the wheel. One human being could stand on the technological shoulders of another. With the gradual accumulation of knowledge, civilizations would emerge. The monuments of ancient Egypt still symbolize the glory of these ancient civilizations and their inventions.

For most of human history technical progress has occurred very slowly. Often it has even contracted. Fifteenth-century China was more technologically advanced than nineteenth-century China. For almost fifteen hundred years, in what we know as the Dark Ages, European per capita incomes were below the peaks achieved during the Roman Era. But even in these dark times technical progress did not stop—the human animal is inventive and its inventiveness cannot be stopped. The compass, the rudder, gunpowder, and the horse collar appeared in this era. But social disorganization meant that newly developed technologies spread slowly and that old technologies were little used. Ancient Romans knew how to make and use chemical fertilizers. The farmers of the Dark Ages did not. Crop yields per hectare fell dramatically. Cities shrank and disappeared. At the depth of the Dark Ages, the largest cities in Europe were less than one percent as large as Imperial Rome at its peak. Literacy rates plunged.

Over time, technological leadership has moved from one part of the world to another. In ancient times it was found in Egypt and the Euphrates River Valley, then it moved on to Greece and Rome. During Europe's Dark Ages, the Western center of learning and innovation moved to the Islamic world, where modern numbers were invented.

In the fifteenth century, before Columbus landed in America, China was the world's technology leader. It produced more iron and steel than eighteenth-century Europe would. China had gunpowder, the compass, and the rudder. It had put an armada with 28,000 troops on the east coast of Africa with ships four times as big as those of Columbus. It had moveable type and paper. It had the plow and threshing machine. And, believe it or not, it knew how to drill for natural gas. To this day there is a debate as to how many "European" inventions are really technologies that slowly moved from China, where they had been invented much earlier.

But the long European downturn eventually ended. The Renaissance began. In the fifteenth century the printing press (probably invented twice—once in Germany and earlier in China) made it feasible for everyone to participate in the cultural revolution that we know as literacy and numeracy. Books became cheap; it was no longer just the elite who could afford reading materials. The door was open for everyone to learn what others had already learned.

Most of what we think of as technical progress has happened in the last 200 years. With the invention, or more accurately perfection, of the steam engine (ancient Egyptian temple doors were opened and shut with the pressure of steam), the industrial age arrived. What was developed as a device to pump water out of coal mines caused a revolution in transportation (the railroad) and powered the development of a system of industrial factories (first in textiles and then elsewhere).

Eight thousand years of agriculture as the dominant human economic activity had come to an end. Ninety-eight percent of Americans worked on the farm in 1800. Today only two percent are full-time farmers.

If one wanted to be rich as an individual, company, or nation, one had to play the industrial game. Prior to the steam engine, differences in per capita income between countries were small. Everyone depended upon agriculture for their wealth. But with the first industrial revolution it was possible to raise per capita Gross Domestic Products far above the levels afforded by agriculture.

Some countries made that leap into the industrial age. Others did not. What we call the first world and the third world came into existence. Everyone knew about the new technologies, but some countries had the social organization to leap into this new world and some did not. Two hundred years later some countries are still struggling to make it into the industrial age.

At the end of the nineteenth century one great idea and one great discovery changed the nature of technical progress. This great discovery was that electricity could be harnessed (with lightning and static electricity, the existence of electricity had been known for a long time) and used. Electricity replaced lamp oil in lighting, the steam engine in transportation—and allowed the telegraph to be invented. Communication at the speed of horse and rider (the pony express) was replaced by communication at the speed of light. At the beginning of the twenty-first century it is impossible even to conceive of modern life without electricity.

The great idea (a German idea that arose in the course of developing their chemical engineering industries at the end of the nineteenth century) was that systematic investments in research and development based upon academic science could lead to much faster rates of technological progress. In the nineteenth century the world's technological leaders were to be found in Great Britain. Today we would call them great industrial tinkerers. Bessemer, for example, never knew—chemically speaking—what was going on inside his blast furnaces as he made steel. He just found by experimentation what worked and what did not work. By contrast, in the twentieth century most inventions sprang from systematic scientific investments in research and development (R&D).

The industrial research and development laboratory, our great research universities, government research budgets (the National Science Foundation, the National Institutes of Health), and government laboratories all flowed from this great German idea. Invention still depended upon creative individuals, but the scientific platforms upon which they worked had to be socially created and supported. Technical progress had been systematized and was no longer just the product of fortunate accidents. It came to be seen as the normal condition of human kind. To the ancient Egyptians, Greeks, or Romans, forever advancing technical progress—something we all assume to be true—would have been a concept impossible to comprehend.

Based upon systematic investments in R&D, Germany became the world's technological leader in the first half of the twentieth century. Its scientists, or those closely related (Werner Heisenberg, Niels Bohr, Albert Einstein), and its inventions (V-2 rockets during World War II) were ahead of those in the rest of the world. Until the 1940s, if one wanted the best scientific education, one went to Germany. World War II led to the eclipse of German science and the rise of American science. To a great extent this shift was led by the brilliant individuals who moved from Europe to the United States (such as Einstein and Enrico Fermi).

Today scientific understanding, engineering advances, invention, innovation (bringing a product to market), and the widespread use of new products in human life are seen as a logical, linear process. Sometimes they are: Einstein's theory of the equivalence of mass and energy, $E=mc^2$, led to the development of atomic weapons. But often they are not. Invention often comes before scientific understanding, and this leads to the search for the principles that make those inventions possible. Steel was made long before we understood the scientific principles of how iron became steel. Selective breeding of plants (replanting the seeds of the most prolific, hardiest plants) occurred long before genetics was understood.

What seems useless is often found to be useful. Often this depends on complementary technological developments. Bell Laboratories' lawyers did not want to patent the laser because they could think of no uses for it. Today, nothing is more ubiquitous than the laser—for correcting your eyesight, playing your music, and transmitting your telephone calls.

In the end, invention is a mysterious process. It almost always involves hard work and countless dead ends. Luck often plays a role. But in the end the "Eureka!" of the inventor reflects reality. Something not known, or not possible, has become both known and possible.

Inventions and innovations do not occur evenly in human history. Long periods with little activity are interspersed with periods of intense technological creativity. Historians see these latter periods as technological revolutions, as with the agricultural revolution 8,000 years ago, the steam revolution at the end of the eighteenth century, and the electrification revolution at the end of the nineteenth century. Today we are in the midst of what I believe will come to be called the third industrial revolution. This revolution is based upon rapid advances in six technologies—microelectronics, computers, telecommunications, new man-made materials, robotics, and biotechnology—and the interactions between them.

The interactions are probably even more important than the leaps. When physics and microelectronics met biology, for example, robotic microelectronic analyzers allowed the sequencing of the human genome to be completed six years earlier than expected. At the same time, biology was informing physics and electrical engineering: for example, researchers began building computers based on molecular, not electronic, principles.

Whenever there is a profound change in the way we understand the world—and biotechnology should be seen as the Newtonian and Einstein revolutions in physics compressed into a 50-year period—there is a period of ferment, invention, and innovation. Old production modes are suddenly replaced with new ones (e-commerce replaces the neighborhood store). Never-before-seen technologies (the Internet) change how humans live their lives.

In the short run, technical progress and economic progress are not identical. The deployment of knowledge requires social organization, and often that key ingredient is absent. That is why Imperial Rome had a much higher level of economic activity and standard of living than the Dark Ages, even though the Dark Ages had technologies not available to the Romans.

But in the long run, technical progress and economic progress move on parallel tracks. Without technical progress, even in a perfectly organized society, economic progress would quickly stagnate. All of the investments that could profitably be made would have been made and there would be nothing new in which to invest. Improvements in standards of living would come to a halt.

This book celebrates the technical advancements that make human progress possible. It presents snapshots of the twentieth-century innovation that led us to where we are, and it is a peek into the torrent of inventiveness that is appearing at the start of the twenty-first century.

These stories do not look at the whole process. They look at inventions and not the scientific discoveries that made those inventions possible. This is not a book about Nobel Prize winners. The book also does not focus on innovation—the hard process of bringing a product to market and making it into something that humans want to use. Persuading Americans to use a garbage disposal was not easy (how could one put garbage in the sink where one was washing one's dishes?). In fact, it took the dishwasher to make the garbage disposal a common household appliance (dishes were now not washed where garbage was deposited).

Some inventors are innovators. Ford invented the assembly line and built a company and a product based upon what he invented. Most inventors, however, are not. Philo Farnsworth invented the television but no company or TV set bears his name. Conversely, most innovators are not inventors. Carnegie built America's steel industry but he did not invent the technologies that he used.

This book springs out of the Lemelson-MIT Program, based at the Massachusetts Institute of Technology, which every year awards the world's largest single prize for invention, the $500,000 Lemelson-MIT Prize, to an American inventor.

The aim of this prize (inspired and financed by Jerome and Dorothy Lemelson nearly a decade ago) is to bang the drum for American invention and to provide role models for young Americans. They, too, might become inventors when they grow up.

Many of the world's greatest inventions were not American—the jet engine came from Great Britain, the printing press from Germany, the wheelbarrow from China, the World Wide Web from Europe. This book focuses on American inventors, not because the country has a monopoly on inventing, but because it is designed to help Americans understand the process of invention and to excite and stimulate them to become more inventive.

At the same time, it is worth focusing on American inventiveness at the turn of the twenty-first century. Other nations now look to the American model to find ways to increase their inventiveness. It explains why the European and Japanese newspapers that were writing about the end of the American century a decade ago are today writing about the difficulties of keeping up with new American technologies. The technological revolution now under way is not the whole answer to this turnaround in American economic success (American firms did some very rapid learning about total quality management, just-in-time inventories, and design for manufacturability from the Japanese in the late 1980s and early 1990s), but it is a large part of the answer.

Perhaps there is another justification. Americans have not always been the world's technological leaders. In the nineteenth century that honor was held by Great Britain, and in the first half of the twentieth century it was held by Germany. Technology and American leadership have only come together in the last 50 years. What has been gained can be lost. Understanding how those who gave us that leadership came to be creative can perhaps help us keep the creativity going far into the twenty-first century.

The inventions and inventors in this book were picked both because they have contributed to raising human standards of living and because they are interesting—good fun, if you like. Some of them have contributed to making human life longer and healthier—medical devices, water purification, the traffic signal. Some of them helped make life more fun—TV, video games, the outboard engine. Some are ubiquitous—plastics, the computer mouse. Some serve the human desire to wander—the auto assembly line. But they have all contributed to making us what we are.

This book is first and foremost a tribute to American ingenuity and inventiveness. Second, it may help us understand the conditions under which ingenuity occurs. Third, it is designed to provide role models for tomorrow's inventors; wishing to become Steve Wozniak (inventor of the Apple computer) is no more outlandish than wishing to become Babe Ruth. Fourth, it is a natural outcome of the invention outreach initiatives of the Lemelson-MIT Program.

This book is not meant to be a comprehensive listing of twentieth-century American inventors, and it certainly does not reflect our judgment as to who are the 35 most important inventors. They were selected because they have made important contributions and because their stories are interesting; they are inspiring individuals. Some have been picked because we hope they will surprise the reader. Their stories are ones you don't already know.

The book includes profiles of the inventors, organized into five main categories: medicine and healthcare, consumer products, transportation, energy and the environment, and computing and telecommunications. Science historian James Burke has written an introduction to each section. Burke's introductions are designed to lay the groundwork and provide a historical context for each section.

We want readers to have fun with this book. We want them to say, "I could do something like that too." We want readers to set out on great voyages of intellectual discovery themselves. And just as those who set out on the great voyages of geographic discovery aren't always remembered in history, most of the inventors in these pages won't be found in the history books. But these inventors have had courage and perserverance not to discover, but rather to invent, a new world. We hope you enjoy the journey.

To journey further on line, visit:
www.inventingmodernamerica.com

INVENTING MODERN AMERICA

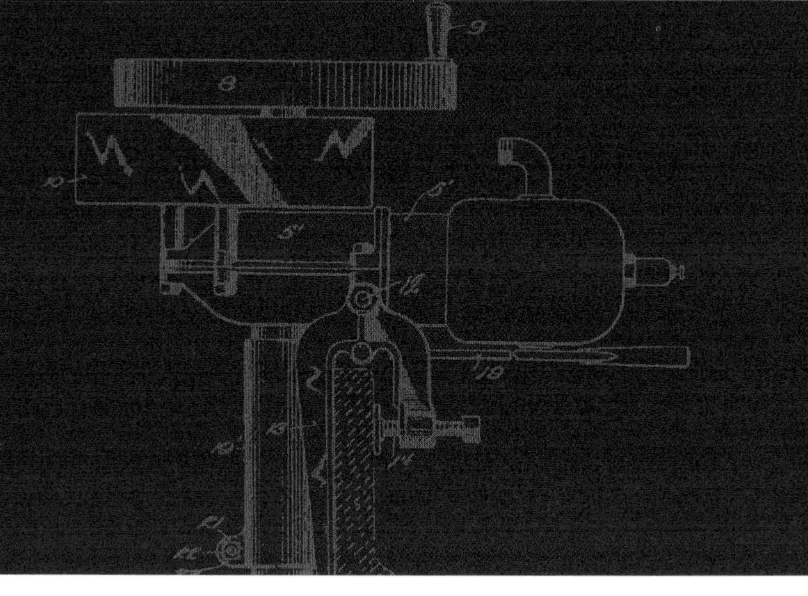

INTRODUCTION

James Burke

It is ironic that the United States should be the most technologically innovative country in the world, with an annual patent rate running at about 70,000 (not to mention the hundreds of thousands of unpatented inventions)—and more than six million patents issued since the Founding Fathers opened the U.S. Patent Office in 1790. The irony is that it was the rediscovery of America by Columbus that triggered it all.

The sixteenth-century canonical authorities (the Bible and Aristotle) made no mention of the new continent, so what was it doing there? Over the following hundred years the avalanche of new plants and animals (such as pineapples and turkeys and potatoes) flooding into Europe from the New World (and none of these had been included in the canonical lists, either) brought a crisis of intellectual confidence. If the classical Greek knowledge on which everyone relied was wrong, then what was right? Things became even more perturbed in the sixteenth century when Copernicus's *De Revolutionibus* turned everything upside down when his heliocentric system removed the Earth from the center of the cosmos. As the contemporary English poet John Donne later understated it: "The new philosophy calls all in doubt."

In an effort to find a more secure way to generate trustworthy knowledge, the seventeenth-century French engineer René Descartes produced two rules: Doubt everything that is not self-evident, and reduce all problems and structures to their simplest, irreducible components, so as to understand how they work.

Methodical doubt and reductionism laid the foundation of scientific thought and fail-proof technological practice, and triggered four centuries of innovation that modern inventors such as those in this book can now draw on. By the early twenty-first century, Cartesian thought had generated hundreds of scientific and technological disciplines, each working according to natural laws discovered by the same Cartesian process. Before Descartes we had lived by the simple axiom *credo ut intelligam* (faith brings comprehension). Descartes reversed it: *intelligo ut credam* (when I have examined the proposition for traps I'll get back to you).

As any scientist will agree, reductionism requires unremitting attention to detail. Some modern fields of endeavor, such as genetics or particle physics, involve mind-numbing patience and months of routine analysis before even the smallest advance can be achieved. Thomas Edison, the inventor's inventor, defined genius as "99 percent perspiration and 1 percent inspiration." He was speaking from experience. To arrive at the right filament for his light bulb Edison and his trusty assistants tested thousands of materials over many months. Recognizing this requirement for exhaustive investigation, Edison systematized the process with the world's first research lab in his Menlo Park, New Jersey, establishment, and brought the innovative production technique down to a fine art: a minor invention every ten days and a "big trick" every six months or so. A colleague of Edison's, Nikola Tesla, summed him up: "If Edison had a needle to find in a haystack he would proceed at once with the diligence of a bee to examine straw after straw until he found it."

Thanks to the kind of human ingenuity celebrated in this book, innovation has brought us to the threshold of transition from a culture of scarcity to one of abundance, in which we may have the technology to release the awesome potential of more than six billion imaginative brains around the planet. Each of those brains is adept at (perhaps even designed for) thinking in the fundamentally innovative mode celebrated in this book. The kind of thinking that causes one and one to make three: an evolutionarily valuable characteristic, since it permits the brain to run scenarios, to bring together the various elements of any situation, predict the likely outcome, and make a decision based on that prediction.

All of the inventors in this book do this, in some way or another. And, while in no way detracting from their brilliance, or that of the thousands of innovative minds that preceded them both in America and elsewhere, perhaps their greatest contribution has been to develop the tools and provide the opportunity for any other human brain to do the kind of innovative thinking they have done.

We can all be all inventors, just like the ones in this book. They show the way.

MEDICINE AND HEALTHCARE

INTRODUCTION

The first group of inventors featured in this book displays the same talents for attention to detail as the early reductionists. They all work in the world of medicine. Not surprisingly, reductionism has brought astonishing developments in this (literally) life-and-death field of study. Medicine, as our inventors' biographies show, has a history of dogged application mixed with moments of wonderful serendipity. In 1928 Alexander Fleming returned from holiday to his London hospital to find a mold growing on the unwashed lab dishes he had left stacked in the sink. The mold spores had come in through a window accidentally left open, the air had stayed at exactly the right temperature and humidity for the right number of days for the mold to grow, and the lab cleaner happened to have not washed up. This combination of events made Fleming's discovery of penicillin possible.

Similar periods of application interrupted by serendipitous flashes have enlivened the work of our first group of inventors. On one occasion, a patch of seaweed inspired a novel method for growing artificial human tissue. In the case of Mary-Claire King, her discovery of the marker for the gene related to breast cancer came only after years of unremitting step-by-step analysis.

In every case, there was also a defining moment of insight, when all of the inventors say they suddenly saw the answer to the problem. Each time, the innovative moment came when they were able to look at things in an unconventional way. Thomas Fogarty unexpectedly understood the relevance to his cardiovascular work of a technique he often used to tie fishing flies. Raymond Damadian's MRI scanner came out of a realization that a way of using a magnetic field to analyze physical substances could be employed to provide pictures of the inside of a living human body.

The other common characteristic of this group is (naturally enough, given their specialty) an abiding desire to help people. As twenty-first-century American science and technology offers greater and greater insight into the workings of the human body, opportunities will proliferate for minds like these to use the widening pool of knowledge to help them achieve the goal they share: to better the human condition.

—**JB**

FANTASTIC SUCCESS!
4:45 A.M. First Human Image Complete in Amazing Detail

RAYMOND DAMADIAN

At some point in June of 1977, the huge, heavy, complicated machine that Dr. Raymond Damadian and his assistants were building needed a nickname. Since the seven years they had spent building and rebuilding it had been full of struggle, adversity, and near-failure, Damadian gave it a name that reflected the courage and dedication that went into it: Indomitable. Less than a month later, the machine was switched on and, over the course of five hours, it delivered the first magnetic resonance image (MRI) of a living human.

That first MRI, a relatively crude picture of the chest of one of Damadian's assistants, ushered in a new era in medical imaging. Today, MRI machines are used at every hospital to produce safe, noninvasive images that detect cancer and other diseases, guide surgeons, and reveal some of the body's secrets. Indomitable, the machine that started it all, has found a permanent home in the collection of the Smithsonian Institution.

The story of Damadian's invention begins with his research into one of those secrets. As a postdoctoral student in the 1960s, he was trying to find out how the kidney moved sodium it had absorbed back into the body. Most people assumed that there was a "sodium pump" in the kidney, but no one had found it or come up with an alternative explanation for the process. Damadian searched for a pump for five years before concluding that one did not exist. So he began reexamining how the body's cells work, specifically how water helps sodium and potassium move into and out of them. In 1969, he was discussing his research with a physicist and physician, Freeman Cope, who had measured sodium in cells with a device called a nuclear magnetic resonance (NMR) spectrometer, and hoped to measure potassium. Damadian was intrigued by the method and offered to help.

The phenomenon of nuclear magnetic resonance (NMR) had been discovered in 1937 by the physicist Isidor Rabi. His research, later refined by physicists Felix Bloch and Edward Purcell, revealed a fundamental property of physics. When radio waves were aimed at atomic nuclei in a strong magnetic field and then turned off, the nuclei emitted consistent, measurable amounts of energy for a short period of time. That period of time—called the relaxation

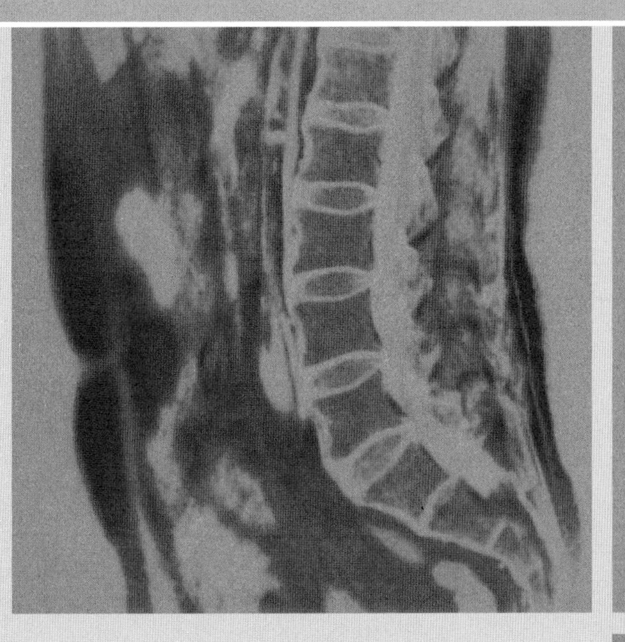

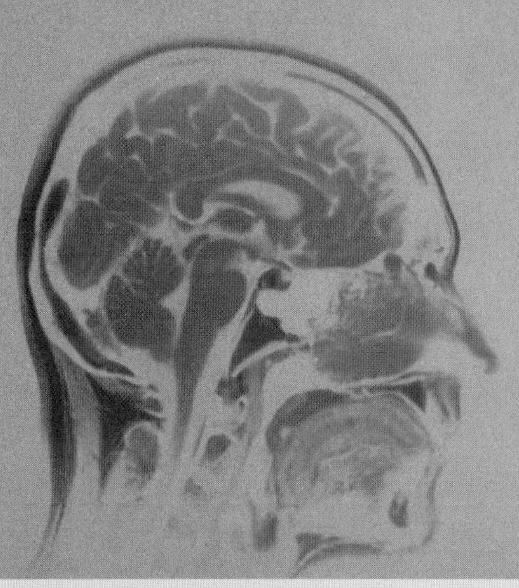

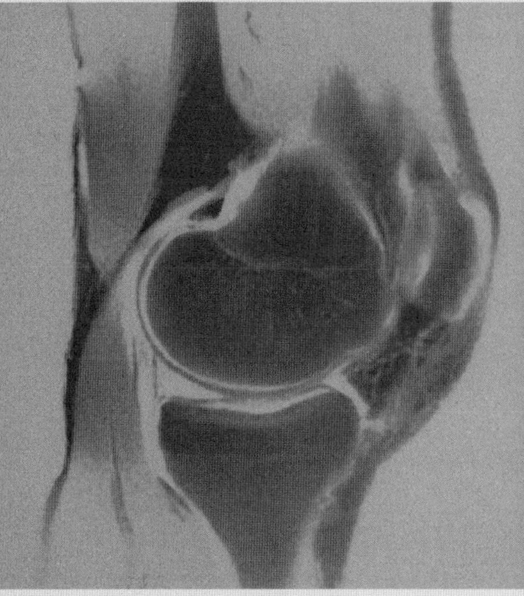

Originally conceived as a method for detecting cancer, the MRI has become the standard for diagnostic imaging, producing pictures of the body's internal structure and organs with unparalleled detail.

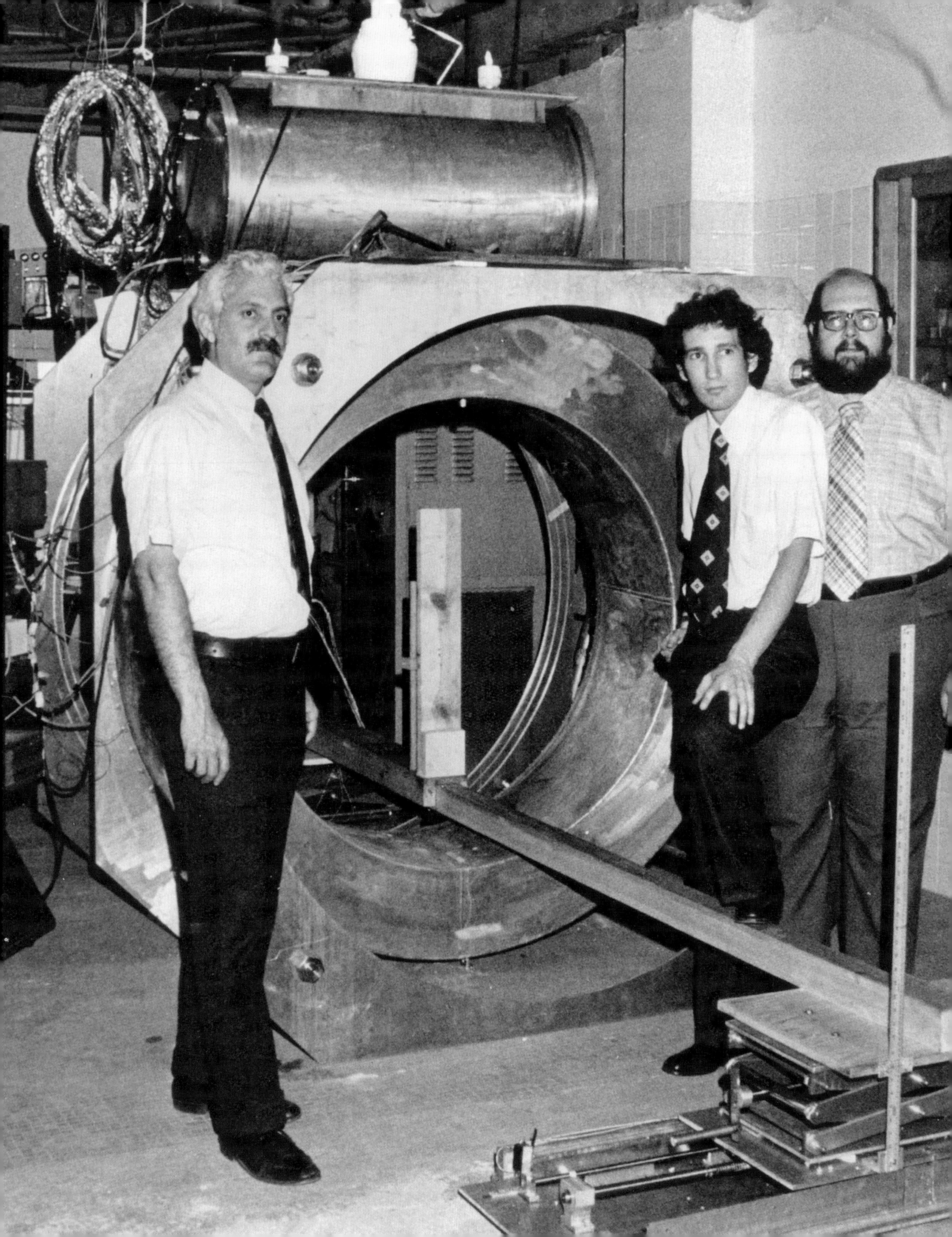

rate—differed for different substances. Soon scientists were using NMR to determine the physical properties of all kinds of things. Until Damadian and Cope, though, no one had used NMR to probe the workings of living cells.

The two scientists traveled to a NMR lab in Pennsylvania to examine a potassium-rich strain of bacteria. They inserted a sample into the NMR machine—essentially a large magnet equipped with an antenna that emitted radio waves—and saw an oscilloscope signal that revealed the bacteria's potassium levels. Damadian was awed by this result, something he could barely have replicated in a lab. "It had a profound effect on me," he said. "It was doing chemistry by wireless electronics."

A few days later Damadian had a vision of where NMR could lead. "If you could ever get this technology to provide the chemistry of the human body the way it does for the chemist," he said, "you could spark an unprecedented revolution in medicine." For one thing, Damadian realized that cancerous cells have a different proportion of sodium and potassium than healthy ones. If NMR could detect that difference—and he thought it could—then a doctor could diagnose cancer early and without surgery.

In June 1970, Damadian took the first step in doing just that. Working with healthy and cancerous tissue samples from rats, he successfully used NMR to detect which was which. He also realized that each type of tissue, whether healthy or not, had a different relaxation rate, so NMR scanning was not limited to detecting disease—it could provide a basic window into the body. The next year, he published his findings in the journal *Science*, and in 1972 filed for a patent. The patent was awarded in 1974, the first covering the technology.

There was a personal note to his quest for a successful imaging technique: For several years in the 1960s, Damadian had been plagued with abdominal pain. After several doctors failed to find its cause, Damadian decided that there should—and could—be a better way to examine the body. Also, as a child he had witnessed his grandmother succumb to a slow and painful battle with breast cancer, inspiring a lifelong quest to conquer the disease.

Born in 1936 in New York, Damadian grew up in Forest Hills, Queens, surrounded by high-achieving peers. He was a prodigy at the violin, attended the Juilliard School of Music, and dreamed of becoming a solo violinist. At 15, Damadian realized that very few people are good enough to be professional musicians, so he gave up studying music and accepted a scholarship to the University of Wisconsin. He soon decided to become a doctor, and in 1960 graduated from the Albert Einstein College of Medicine in the Bronx. The combination of science and medicine—as well as a fierce fighting spirit picked up as a tennis player—would serve him well.

The 1971 article in *Science* should have been Damadian's finest hour, an announcement of a new way of seeing the body. Instead, it was the beginning of years of dismissal and ridicule by his peers, as well as cutthroat competition. His ideas of using technology from the world of physics to probe the body were completely new, and his long-range vision of using NMR to detect disease struck many scientists as wild and improbable.

Raymond Damadian, Larry Minkoff, and Michael Goldsmith stand next to their just-completed magnetic resonance imaging (MRI) machine, called Indomitable.

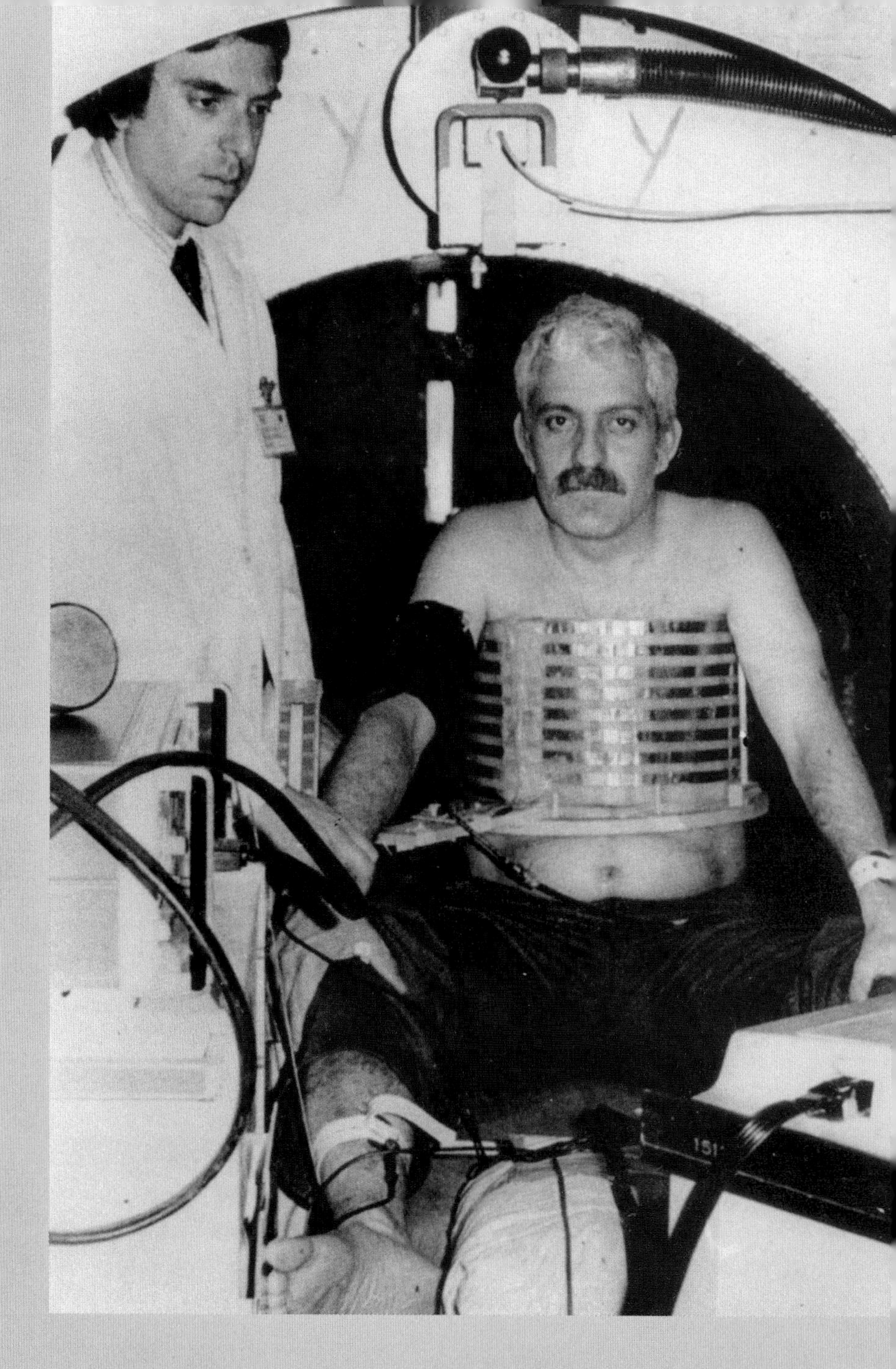

In May 1977, Damadian wrapped himself in a cardboard antenna and became the first test subject of the first MRI machine. The test was unsuccessful; the inventor was too big for the antenna.

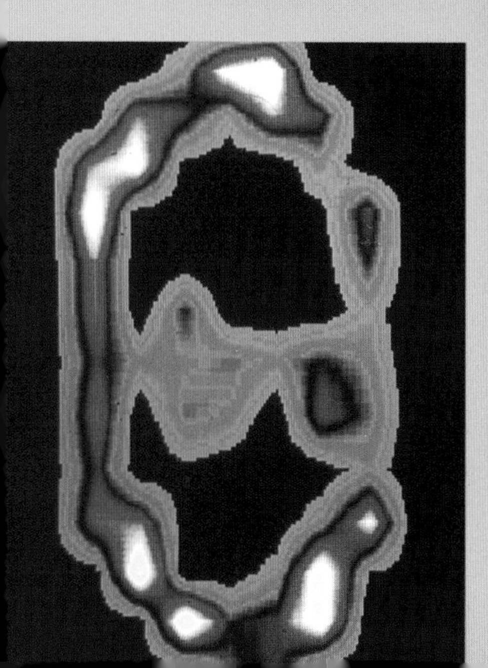

On July 2, 1977, the slimmer Larry Minkoff entered Indomitable. For five hours, Damadian and Michael Goldsmith recorded the machine's readings; they later created this digital image of Minkoff's chest from the data.

In his lab at the Downstate Medical Center in Brooklyn, Damadian assembled a small group of assistants and began exploring how to make a machine that could use NMR to scan a human. Several challenges faced them. First, they needed to build a magnet big enough for a human to fit in; only a handful of magnets that large existed, for use in nuclear physics research. They also had to find a way to focus the NMR and move it over a human body; the machine Damadian had used in his cancer cell study was designed to analyze small quantities of a single substance, so it didn't need a sharp focus or have to move.

Working on a shoestring budget, Damadian and his assistants built a large superconducting magnet and its liquid helium cooling system, often working into the early morning winding miles of wire and searching for helium leaks. He devised a method for focusing the NMR's radio waves, so as a human was moved through the device, different parts of the body could be seen. Since no one had built a large, focused NMR device before, the team constantly had to break new technical ground, experimenting to find the best way to make almost every part of the machine.

Through much of the 1970s, Damadian's main competition came from Paul Lauterbur, a chemist at the State University of New York in Stony Brook. In 1973, Lauterbur used NMR to scan the interior of a small clam—the first living creature to be imaged. He also developed a different way to focus the machine's signal, a method that gained wide use. Although Damadian thought of him as his rival, the two eventually shared the National Medal of Technology (1988) for their achievements.

In May 1977, Damadian thought Indomitable was ready to be tested. He volunteered himself as a subject, although no one knew what effects were possible from prolonged exposure to a strong magnetic field. Wrapped in a cardboard antenna, he lay down in the machine as it was switched on. But there was no NMR signal from his body; he was too big for the device. Two months later, on July 2, 1977, his slimmer assistant, Larry Minkoff, entered Indomitable. Damadian and his colleagues slowly read NMR signals from different levels of Minkoff's chest, plotting their values on a piece of graph paper. After almost five hours, Minkoff emerged to see a cross-section of his own chest, drawn in pencil. (A computer would be used later to replot and enhance the image.) It was the first scan of a human.

Three years later, Damadian's company, Fonar, built the first commercial MRI scanner. (Because of the negative publicity then surrounding all things nuclear, "nuclear magnetic resonance" was shortened to "magnetic resonance," hence MRI.) Scan time had been reduced considerably, and the image's resolution was many times higher than Indomitable's first effort. While Damadian had met with intense criticism for years, MRI machines became common (and soon essential) features of every good hospital.

Success breeds competition, and Damadian's small company soon found itself out-spent and out-marketed by Hitachi, Philips, and General Electric, among others. By the 1990s, Fonar had only a one percent market share. But Damadian is a firm believer in the power of the U.S. patent system, and he began to challenge other MRI makers. Many companies settled out of court, but GE—which had rejected his ideas in the 1970s, then began working on its own version of a scanner—fought his patent. In 1997, the Supreme Court upheld a judgment against GE, ordering it to pay Damadian damages of $128.7 million. After almost three decades of effort, his pioneering work had been validated not only in hospitals around the world, but also in the nation's highest court.

THOMAS **FOGARTY**

Thomas Fogarty entered medicine through the back door, or at least the supply room door. As an eighth-grader, he got a job dispensing supplies at Good Samaritan Hospital in Cincinnati, Ohio, and worked his way up to being a scrub technician, the person who hands a surgeon tools during an operation. As Fogarty observed the perils of surgery, especially surgery to remove blood clots, his inventive mind explored ways to improve operations. While still a scrub tech, he conceived and built the balloon catheter, a tool to remove clots without major surgery. Since its first use in 1961, the balloon catheter has saved or improved the lives of millions of patients.

The balloon catheter also marked the beginning of noninvasive surgery. By minimizing trauma to the patient and reducing time spent under anesthesia, surgical procedures such as blood clot removal became much safer and more cost-effective. Noninvasive procedures are now the standard of care for many health problems that once required lengthy surgery.

Fogarty showed his inventiveness at an early age. He was born in Cincinnati in 1934; his father, a railroad engineer, died when Fogarty was just eight years old. Fogarty took an active role in the family, taking care of things his father might have. "If my mother needed things fixed," he remembers, "she would call on me." It wasn't much of a stretch for Fogarty: "I just had a natural inclination and inquisitive nature about building things," he says. "I looked at things and just naturally thought, 'Okay, how can I make this better?'" His first mechanical explorations were with soapbox derby cars and model airplanes. Fogarty's planes were so good that he sold them to other kids in the neighborhood. He eventually graduated to more complex machines, devising, for example, an automatic clutch for a friend's motor scooter.

With money tight in his family, Fogarty went to work when he was 14 years old. Devoting time to work and mechanical tinkering, and having little interest in his classes, he did not do well in high school. However, by the time he was a senior, Fogarty knew he wanted to be a doctor. With the recommendation of a family priest, he was admitted to Xavier College in Cincinnati, although

Thomas Fogarty's inventiveness first surfaced in the soapbox derby cars he designed and raced in Cincinnati in the 1940s. He soon moved on to modifying model airplane engines and motorized scooters. He is shown here in his "Flying Irishman," a go-cart that he built.

Fogarty's first medical device was the balloon embolectomy catheter—which he made by tying a small balloon to the end of a urethral catheter. It is inserted into a blocked blood vessel and then inflated and pulled back out, bringing a clot, or embolism, with it.

he was immediately put on academic probation. (He quickly learned how to study.) Fogarty needed three jobs and the financial help of a mentor, Dr. Jack Cranley, to get through college. He kept his scrub technician job and also worked as an X-ray technician and an orderly on the night shift.

It was as a scrub tech during college that Fogarty began to think about better ways to perform various surgical procedures. "I had a chance to see what worked and what didn't work," he says. "One of the things that didn't work was the way they took care of blockages in arteries and veins." At the time, the procedure for removing a clot was dangerous and not particularly successful. "They would make incisions that were horrendously long and open up the whole artery to try to take the clot out," he says. "That would end up with a 9- to 12-hour operation, with incisions in both legs and the abdomen. And it often didn't work, patients died or they had amputations."

Fogarty considered different ways to make the procedure better—especially how to avoid the long incisions. His first idea was to use a modified catheter. Inserted through a small incision, a catheter could get to the clot without much trauma to the patient. And Fogarty had an idea of how to get the clot out once the catheter was there.

Fogarty's scheme was straightforward. He started with a urethral catheter, which is flexible but strong enough to be pushed through a blood clot. Then he added a small balloon made from a finger of a latex glove, which could be inflated with saline once it was past the clot. The balloon expands to the size of the artery and is then pulled back out, bringing with it the clot.

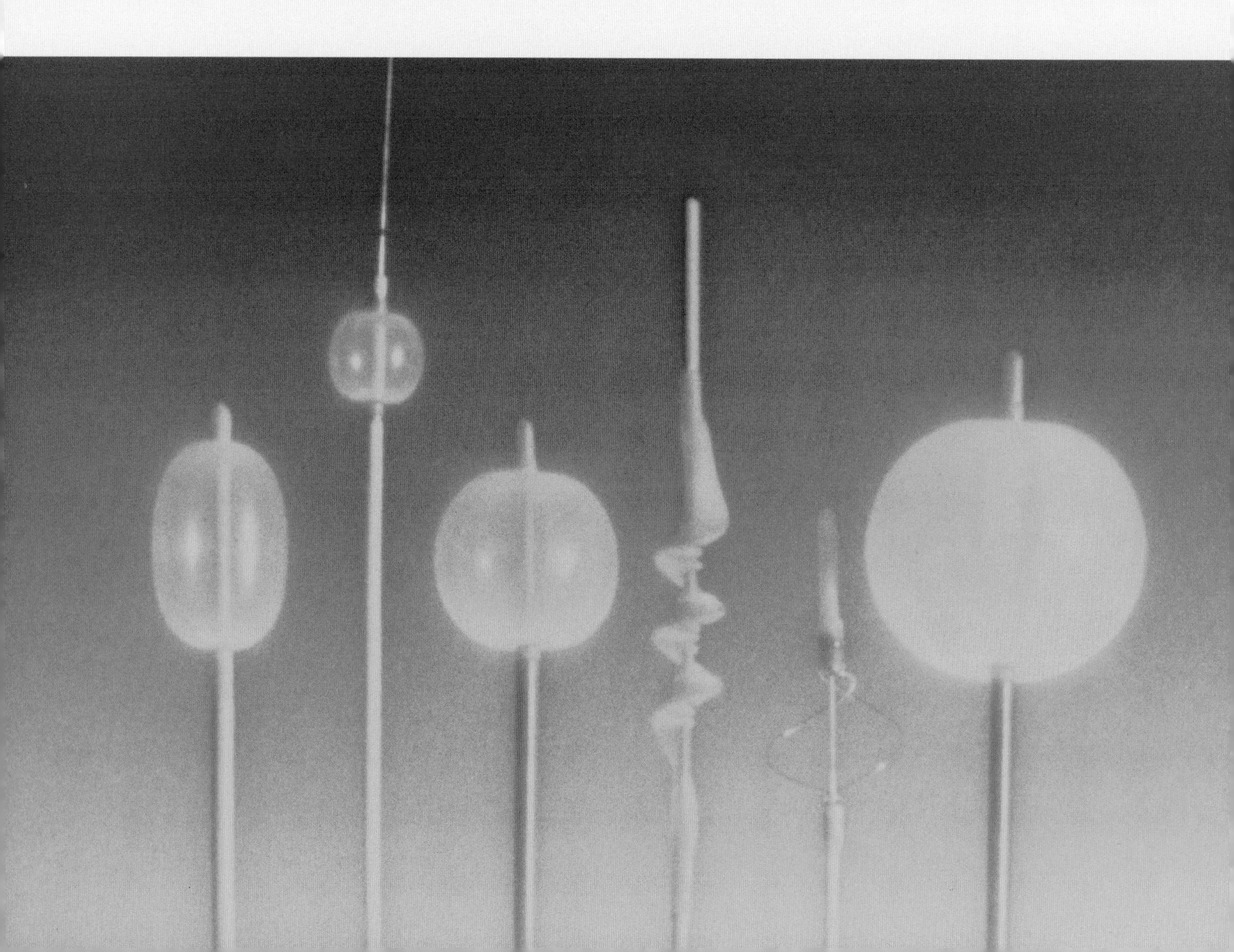

"You can feel the resistance, or the drag on the artery wall," Fogarty explains. "You adjust the volume of the balloon so you've got continual traction."

There was one main challenge in building such a device. "There was no way to attach the balloon," Fogarty says. The catheter was made of vinyl and the balloon of latex, and no glue available would hold the two together. Fortunately, one of Fogarty's hobbies was fly fishing, and he applied the sport's skill of precise hand-tying to the task. "I'd always tied flies and made lures," he remembers, "so it was just a natural thing."

With the balloon attached to the catheter, Fogarty began testing his device. First, he filled glass tubes with blood from a blood bank and let it clot, then inserted his catheter and pulled the clot out. He then tried the catheter on cadavers during autopsies. In 1961, his balloon embolectomy catheter—named for the clot-removal procedure performed—was used for the first time on a human patient. As Fogarty watched, a small incision was made and the catheter was threaded up the patient's blocked artery. When inflated and pulled back out, it did indeed bring the clot with it. Today, the procedure takes about an hour and is done under local anesthesia.

Transforming that single working catheter into the tool that thousands of surgeons now use took several years. When Fogarty was finishing his cardiovascular surgery residency at the University of Oregon, he spent a lot of time making catheters by hand for his colleagues. General acceptance, though, was harder to come by. A paper detailing the catheter's successes in surgery was rejected by three major journals and finally published in a smaller publication's "technique" section. "Traditional wisdom said that this was going to damage the artery," Fogarty explains. "A lot of people would say, 'This is crazy. Only someone that didn't know anything would suggest this.'"

Building catheters by hand was not a practical long-term strategy, but no medical equipment company seemed to be interested in making them. Eventually, Fogarty got to know Al Starr, a cardiac surgeon who had developed the first artificial heart valve used in humans. Starr recognized the catheter's utility and convinced his heart valve manufacturer to make them. Tens of thousands of Fogarty's catheters have been sold since, including several variants for different kinds of clots.

After Fogarty completed his residency and became a cardiovascular surgeon, he continued to invent new medical devices. He holds more than 70 patents, and some 20 companies have been founded to build his devices. One of the most successful of these products is the Fogarty Stent-Graft, an ingenious fix for the difficult problem of abdominal aortic aneurysms. As some people get older, their arteries become clogged with fatty deposits, weakening the vessel. Over time, this can cause the blood vessel to balloon, and eventually rupture. The traditional surgery to fix it, by removing the weakened part of the aorta, is dangerous, with a mortality rate of five percent or more.

Rather than removing the bad part of the vessel, Fogarty's idea was to strengthen it with an implant: a stent, a thin polyester tube that grabs on to the blood vessel with metal "rings." The method begins with an inflatable catheter, which transports the stent to the weakened portion of the vessel. Once it is

Some of Fogarty's catheters: from left, balloon embolectomy catheter, thru lumen embolectomy catheter, venous thrombectomy catheter, adherent clot catheter, graft thrombectomy catheter, and occlusion balloon catheter. Some 20 million people have been treated with these kinds of catheters.

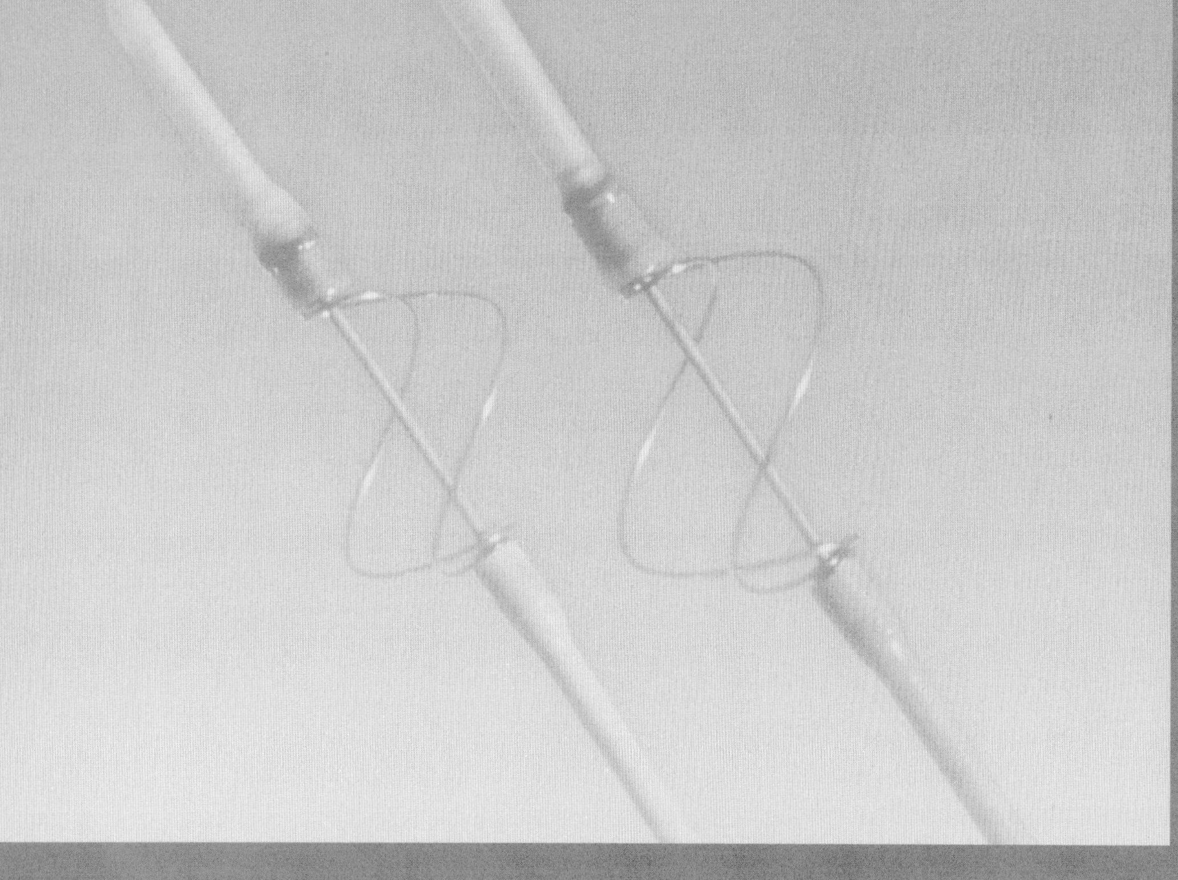

The graft thrombectomy catheter.

“I had no concept that [noninvasive surgery] would reach the magnitude that it has.”

aligned with the help of X-rays, the balloon is inflated and the stent expands to the size of the artery. After the stent is affixed permanently to the artery wall, blood flows normally and the aneurysm is bypassed.

The Fogarty Engineering company is currently developing several new medical devices, including a minimally invasive method for taking biopsies to test for breast cancer. Fogarty himself continues his work as a top cardiovascular surgeon, helps run several other medical equipment and research companies, and presides over the Thomas Fogarty Winery, 325 acres in the coastal hills west of Palo Alto that produce 15,000 bottles of premium wine each year.

When Fogarty built his first balloon catheter, he says, "I had no concept that [noninvasive surgery] would reach the magnitude that it has." Yet the movement toward noninvasive techniques is a logical extension of the concerns he wrestled with back in 1960, as he watched both surgeons and patients struggle with the dangers of blood clot surgery. "If you look at the issues I was trying to solve at that time," he explains, "they are the same issues applied to all these other things we do noninvasively. We minimize the anesthesia, we minimize the tissue trauma, and we shorten the recovery. Those were the things I felt I had to do. And those are the things we now do with everything that's less invasive."

In addition to creating medical inventions, Fogarty is a respected cardiovascular surgeon and also a vintner, operating a small winery in the hills outside of Palo Alto, California.

WILSON GREATBATCH

A tiny mistake led to the invention of the pacemaker, a device that has helped millions of people lead longer and more active lives. In 1958, Wilson Greatbatch was building a small device to record the sounds a heart makes. He reached into a box full of resistors—small electronic components color-coded with their resistance value—and pulled the wrong one out. "I misread the colors and got a brown-black-green [resistor] instead of brown-black-orange," he recalls. The finished device did something odd: It sent out a short pulse of electricity at an interval of one second. This was not useful for capturing heart sounds, but the mistake happened to be exactly what Greatbatch was looking for.

Seven years earlier, while working at the Cornell University Psychology Department's animal behavior farm, measuring sheep's and goats' vital signs, Greatbatch got into a lunchtime conversation with two surgeons. They told him about a condition called heart block, when a nerve fails to send the electrical impulses that makes the heart's muscles work to its lower chambers. This causes an irregular heartbeat and can lead to death. Doctors had already discovered that a quick jolt of electricity to the heart could substitute for the natural process, but the equipment for this was bulky and delivered a painful 100-volt shock to the patient's chest.

When the surgeons described the problem to Greatbatch, he later wrote, "I knew immediately that I could build a much better and smaller implantable device." The problem was that no equipment was small enough for the task; transistors, the small semiconductors that replaced vacuum tubes, weren't widely available yet, and the batteries of the time were too big.

By 1958, things had changed. Transistors had hit the market, and Greatbatch was using one in the small oscillator he was making. When he saw the short pulse of electricity it made, he recalled, "I stared at the thing in disbelief, realizing this was *exactly* the properties of a pacemaker." At the time, Greatbatch was a member of a group of local engineers interested in medical technology.

Wilson Greatbatch's first pacemaker consisted of a simple transistor circuit and 10 batteries. Later pacemakers would have much better batteries, improved casings, and more sophisticated electronics.

In the late 1960s, Greatbatch turned his attention to powering pacemakers and introduced the small lithium batteries that still run millions of the devices. Here he displays an early pacemaker and a schematic of its circuit.

He was called to the local hospital to try to fix a problem a surgeon, Dr. William C. Chardack, was having with some equipment. Greatbatch couldn't help with the problem, but while he was there, he described his idea for a transistor-based pacemaker. "Chardack looked at me strangely and walked up and down the lab a couple of times," Greatbatch wrote, "and then said, 'If you can do that, you can save 10,000 lives a year.'"

Greatbatch finished a prototype pacemaker three weeks later. At the hospital, Chardack and a team of surgeons caused heart block in a dog by tying off the relevant nerve. Then Greatbatch touched the wires of his pacemaker to the slowly beating blocked heart. "Dr. Chardack looked at the pacemaker pattern on the oscilloscope," Greatbatch remembers, "and then said, 'Well I'll be damned!'" The dog's heart was beating normally.

That first working pacemaker was a culmination of Greatbatch's long interest in electronics. Born in 1919, he grew up in modest circumstances in Buffalo, New York, the son of a grocer and a contractor. From an early age, he was fascinated with radio, and in his teens he built a short-wave receiver, then joined the Sea Scouts, partly to use their transmitter. He also used his radio skills in World War II, where he served as a radioman on a destroyer. Later, he was stationed on an aircraft carrier and tended its electronics while also flying combat missions as a rear gunner. After the war, Greatbatch returned to New York and earned a degree in electrical engineering at Cornell, where he was also exposed to other disciplines, including physics and chemistry. Although his education was paid for in part by the G.I. Bill, Greatbatch's finances were tight—he had three of his five children by the time he finished Cornell—so he took jobs with the university. After working on Cornell's radio telescope, he settled into the animal behavior farm, where the idea for the pacemaker was born.

Greatbatch's dog pacemaker was far from the first attempt at a viable, reliable pacemaker for humans. Although Greatbatch didn't know it, several people had already tried. In 1918, an American doctor named Albert Hyman suggested that electrical pulses could shock a heart back to life. In 1933, an electrical engineer, William Kouwenhoven, demonstrated that electric shocks could restore a heart's normal rhythm. In 1952, Paul Zoll, a doctor who had worked with Kouwenhoven, built an external pacemaker; it worked, but not for long, and it was a bulky, awkward device. A few years later, in 1957, Minnesota physician Walton Lillehei and Earl Bakken, a local electronics repairman,

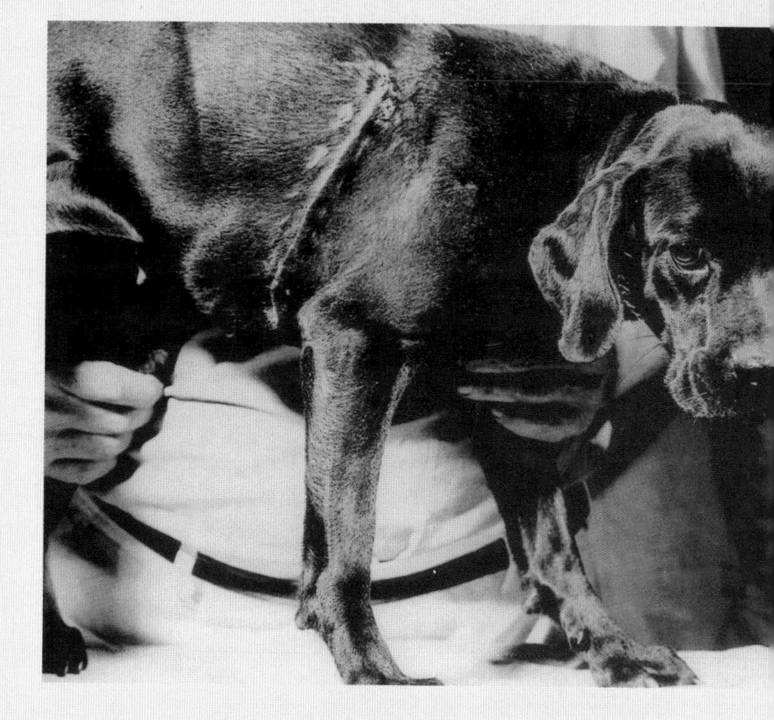

On May 7, 1958, Greatbatch's device was implanted in a dog—and it worked. Later, he wrote, "I seriously doubt if anything I ever do will ever give me the elation—when my electronic design controlled a living heart."

After the success of his first, experimental pacemaker, Greatbatch quit his job and worked to improve the design. He set up this workshop in a barn behind his house; his wife helped build and test the devices, using two ovens set up in their bedroom.

Since the first pacemaker was implanted in a person in 1960, the devices have extended and saved millions of lives. Here, Greatbatch poses with a group of children who have pacemakers.

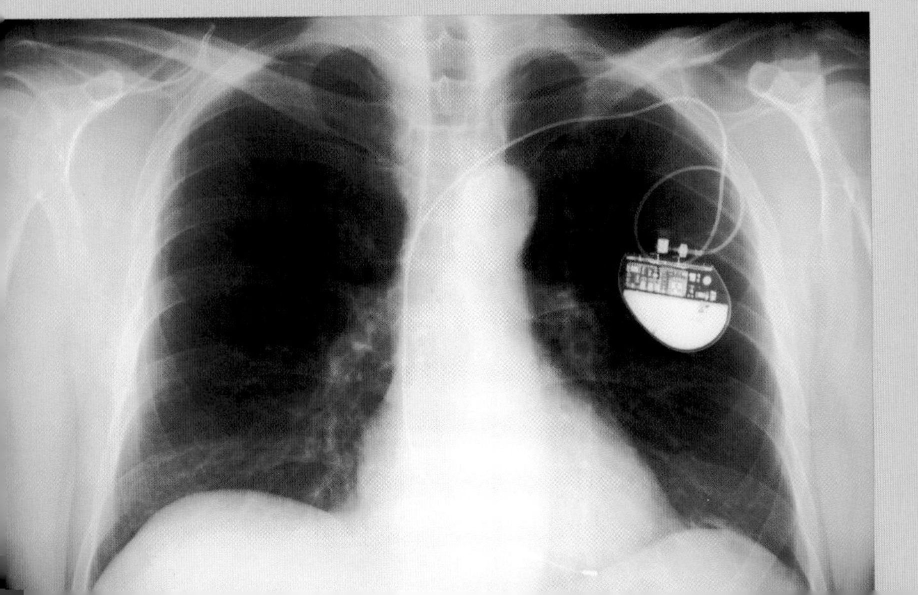

Today's pacemakers contain small but sophisticated computers that monitor the heart and turn the devices on or off or adjusting their impulses as the situation warrants.

built a portable, battery-powered pacemaker, about a foot square. Although their device also worked, its size and the fact that it worked externally made it only a short-term solution to heart problems.

It wasn't just technology that got in the way of the creation of the pacemaker. In the early 1930s Hyman built an "artificial pacemaker" that successfully revived several patients with a condition called Stokes-Adams syndrome. In contrast to the positive reaction to Greatbatch's pacemaker, many people criticized Hyman for intervening in a "natural" death. Others were horrified that doctors would meddle with the heart, the symbol of man's soul.

With the advent of open-heart surgery in the early 1950s, these objections fell by the wayside. After seeing his pacemaker work, Greatbatch set out to make one that could be implanted in humans. He quit his job at Cornell and began working in a barn behind his house, building pacemakers by hand. He hand-soldered the components together—some circuitry, a couple of transistors, and a zinc-mercury battery—and encased the unit in epoxy and silicone. Two ovens were set up in his bedroom to bake the transistors, and his wife, Eleanor, would test each one by tapping it with a pencil. By 1960, he had made about 50 pacemakers, and Dr. Chardack and others started implanting them in human patients. The first recipient lived for 18 months; another led a normal, healthy life into the 1990s.

Those first pacemakers were simple and vulnerable devices. Because of short battery life and the toll exacted by the hostile environment of the body, they usually lasted just two years. In the late 1960s, Greatbatch began working on a new battery. There were two promising kinds: lithium and plutonium. The latter would have lasted as long as a human could live, but the idea was scrapped because plutonium is toxic and radioactive. Greatbatch had greater success with lithium batteries. Not only were they a reliable source of power, they also solved a bigger problem: The earlier mercury-zinc batteries produced a small amount of gas, so the epoxy covering of the pacemaker had to be slightly permeable, which accelerated the decay of the pacemaker. Pacemakers made with gasless lithium batteries could be completely sealed in metal, greatly increasing their durability. Since the 1970s, lithium batteries produced or licensed by Greatbatch's company (which was later run by one of his sons) have powered some 90 percent of all pacemakers. Other pacemaker advances pioneered by Greatbatch and others include units that sense when the heart needs help, and how much.

Throughout his career, Greatbatch, a deeply religious man, strove to uphold a strong ethical code. When he received a patent for the pacemaker, he quickly licensed the technology to another company that could get the devices out to doctors and patients much faster—although the financial return for him was lessened. Greatbatch made it a policy of his various companies to pay for employees' children's college education. And he and his wife have donated millions of dollars to schools and other causes.

Since his work on improving pacemaker batteries, Greatbatch's interests have ranged far and wide. During the oil crisis of the 1970s, he bred a strain of poplar tree that produced a great deal of wood, which he hoped to use as a source of wood alcohol for fuel. He cloned flowers and other plants, and built a solar-powered canoe. In the past decade, Greatbatch has been involved in the fight against AIDS. He and the Cornell scientist John Sanford have received three patents for methods of inhibiting the replication of the AIDS virus and a similar virus found in cats.

Still, Greatbatch's crowning achievement will always be the implantable pacemaker. Dr. Chardack's guess that it could save 10,000 lives a year proved to be an underestimate—more than three million pacemakers have been implanted since 1960. And Greatbatch knew from the moment the dog's heart was shocked into a normal rhythm that something significant had happened. In 1959, he wrote in his lab notebook, "I seriously doubt anything I ever do will ever give me the elation I felt that day when my own two-cubic-inch piece of electronic design controlled a living heart."

DEAN KAMEN

If Dean Kamen had gotten his way as a young man growing up on Long Island, he would have spent his days contemplating esoteric matters. "My hobby was thinking," he says. "It was mostly abstract thinking, trying to understand the universe at a very abstract level." When the financial realities of the world became clear, however, Kamen chose what seemed to him the most obvious path to wealth: He'd become an inventor. Although for many people invention is a surefire way to lose money, not make it, Kamen's creative mind produced medical devices that have been extremely successful and, more important, have improved thousands of people's lives. These inventions include the first portable infusion pump, for delivering precise amounts of drugs to patients; a wearable insulin pump for diabetics; and a portable dialysis machine, which has freed kidney patients from frequent visits to the doctor.

Kamen's realization that abstract thinking wasn't going to turn into a career came to him after graduating from high school. He asked himself, "What am I going to do to make a lot of money?" His answer was, "I could invent something that people want." Kamen's first opportunity came that summer, while he was working for a company that produced audiovisual shows. The equipment the company relied on was clunky and out-of-date; although a new generation of electronics had just come out, it was still controlled with switches and relays. Kamen bought $400 worth of solid-state parts, and over the course of a couple of months he taught himself how they worked, then built a small box that controlled the show in more active and creative ways. He sold the box for $2,000, and then began building them for other companies and museums like the Hayden Planetarium at the American Museum of Natural History in New York City. As he began attending the Worcester Polytechnic Institute (WPI) in Massachusetts, Kamen founded his first company, to make the control systems. By the end of his freshman year, Kamen had sold $60,000 worth of them, all built in his parents' basement.

“The universe is a big playground....”

One of the first things Kamen did with the earnings from his first invention was to outfit his parents' basement as a machine shop. Kamen still does hands-on engineering work in his shop, though it has grown larger and more sophisticated.

A NEW KIND OF MOBILITY

One of Dean Kamen's latest creations harnesses the power of the microprocessor and the gyroscope to solve some of the mobility problems of people who use wheelchairs. Equipped with high-speed microprocessors, a gyroscope, and two independent axles, his Independence 3000 Ibot Transporter, or simply Ibot, can roll up a curb, over sand, and even climb stairs.

Two events inspired Kamen's Ibot. First, he saw a young man struggling to get his wheelchair up a street curb. "It just seemed to me that the fundamental issue was the world has not been architected for people that are sitting down at 39 inches," he told MSNBC's John Hockenberry. The second event happened in his shower—when he stepped out of it one day, he slipped and had to grab a wall to regain his balance. He had already been thinking about how he might use the latest generation of Pentium microprocessors and he now realized he could use their computing power in combination with gyroscopes and small motors to re-create the body's power of balance.

The Ibot can do many things a wheelchair cannot. Its sturdy motor-driven wheels move independently and can raise themselves up to negotiate a curb or stairs. As in a modern all-wheel-drive car, the computer-monitored and -controlled drivetrain automatically adjusts itself to roll over sand, grass, and other irregular surfaces. The Ibot's gyroscopes provide it with a sense of balance; together with its Pentium chips and motors, it reacts to changes in position or weight to keep things stably rolling. Its sense of balance is so good that the Ibot can rear itself up on its back wheels, raising its user to standing height.

Rechargeable batteries keep the Ibot moving all day, and since mechanical or computer failure could be disastrous for its user, backup systems are built in. The Ibot is currently being built and tested by a subsidiary of Johnson & Johnson.

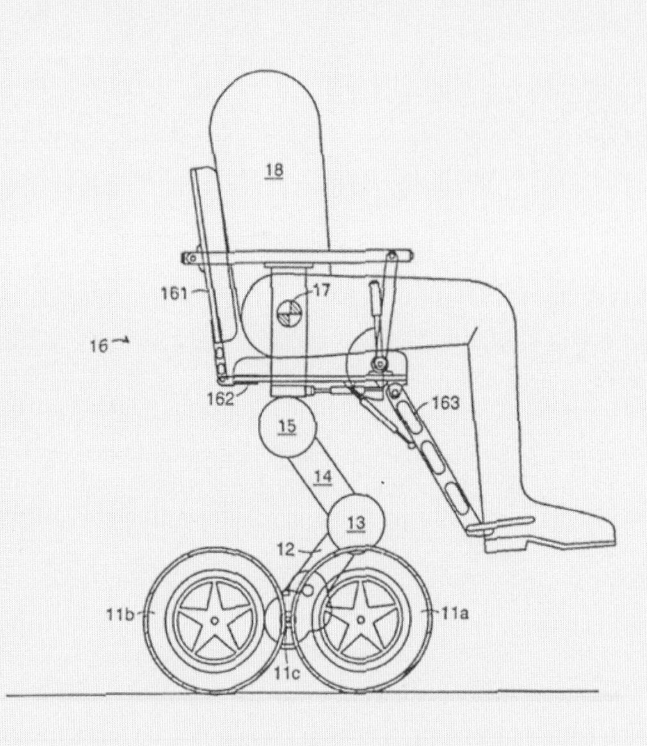

The Ibot, known to the patent office as a "transportation vehicle with stability enhancement."

One of Dean Kamen's most recent inventions is the Independence 3000 Ibot Transporter, or simply the Ibot. Using gyroscopes and microprocessors to help maintain its balance, the Ibot can negotiate stairs and curbs, as well as bring its user to a standing height.

With his earnings, Kamen outfitted the basement with electronics and equipment like lathes and milling machines. At WPI, he was a passionate student of physics but neglected his grades to pursue ideas that interested him. He eventually dropped out.

Inspiration for his next invention came from his brother, who was a medical student at Harvard concentrating on the care of newborns. "He'd come home from med school frustrated," Kamen remembers. "He'd say, 'They have these pumps that are made for adults but I can't deliver microliters of stuff into these babies.'" Building a better pump didn't sound too difficult, so the 20-year-old Kamen got to work. "The conceptual idea of the pump wasn't very hard," he remembers, "but making a device that is going to have lethal, toxic amounts of drug in it that has to work right all the time—that turned out to be more challenging than I thought." A new kind of microchip that required very little power had recently come out and Kamen used it to put together a circuit that controlled a small pump connected to a syringe. The portable pump delivered precisely measured doses of drugs to a patient throughout the day.

A few months later, Kamen's brother showed the pump to a doctor at Harvard, who was immediately excited by it. Between that doctor's enthusiasm and a paper published in the *New England Journal of Medicine*, Kamen was soon flooded with orders. His second business was born, and he put his mother, his younger brother, and his brother's friends to work in the basement workshop, assembling pumps as fast as they could.

Although he enjoyed the technical challenge of making the pump, initially Kamen didn't realize how rewarding his invention could be. "It never occurred to me that people would say, 'Without this my daughter wouldn't be alive,' or 'My mother wouldn't be home,'" he says. "It was *very* gratifying to make things that really helped people." As more and more doctors began ordering his pump, the talkative and inquisitive Kamen built his knowledge of what devices the medical industry needed. "I'd deliver this equipment to different doctors," he remembers, "and my questions to them were always the same: What do you do with this? What else can I do to it? What else do you need? They would always have some unsolved problem."

The next problem that Kamen worked on was delivering insulin to diabetics. When diabetics inject themselves with insulin, their blood sugar can vary widely. Using the same principles as his first pump, Kamen built a portable, wearable device that delivered small, customized quantities of insulin, freeing diabetics from their injection regimen and smoothing out the peaks and valleys of their blood sugar level. The pump was released in 1978.

In 1979, Kamen moved his 20-person company to Manchester, New Hampshire, and three years later he sold it to a large medical equipment manufacturer for tens of millions of dollars. At age 31, Kamen was financially secure and free to do whatever he wanted. He founded another company, DEKA, and developed a new generation of insulin pumps, which are still considered the state of the art. He bought and renovated several nineteenth-century mill buildings on the Manchester riverfront to house the company. And he continued his hobby of collecting antique engines and machinery, some of which are displayed in his office.

Although the engines are often beautiful, his interest in antique technology is not merely aesthetic. For Kamen, there are lessons to be learned from the scientific and technical past. "I think most engineers are worried about how current they can stay," he says. By contrast, one of DEKA's recent products, a kidney dialysis machine that can be used at home, works around physical laws discovered in the seventeenth and eighteenth centuries. (The machine also weighs in at just 22 pounds, one-seventh the bulk of previous dialysis equipment.)

Kamen's current projects include the Ibot, an "advanced mobility system" that could replace the wheelchair. He is also looking at a long-ignored, early-nineteenth-century engine technology, the Stirling Engine, as a way to create small electrical generating systems. The engine, he says, "can be scaled to very small size, it can be used locally, and it doesn't make pollution. Its flexibility in fuel is just awesome." With the electric power industry freshly deregulated, he adds, the engine could "revolutionize the way people think about making and distributing electricity."

In early 2001, news of another Kamen invention caused a sensation. After word of the mystery creation, known as "Ginger" or "IT," was leaked to the press, feverish speculation about what "IT" might be ensued. Many people believe "Ginger" will be a personal scooter powered by hydrogen or a Stirling Engine, and Amazon.com founder Jeff Bezos was quoted as saying it would be "revolutionary." IT, whatever it is, is expected to be unveiled in 2002.

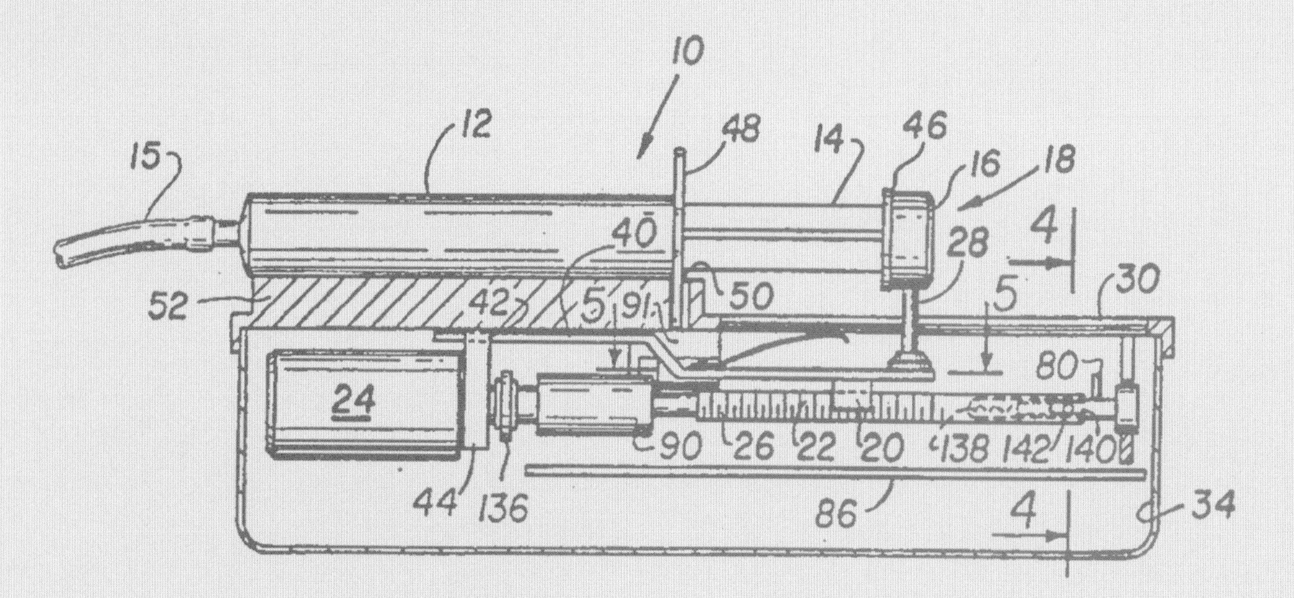

A lightweight kidney dialysis system, patented in 1994.

Kamen's first patent was awarded in 1975 for a portable pump that delivered precise amounts of medication to a patient.

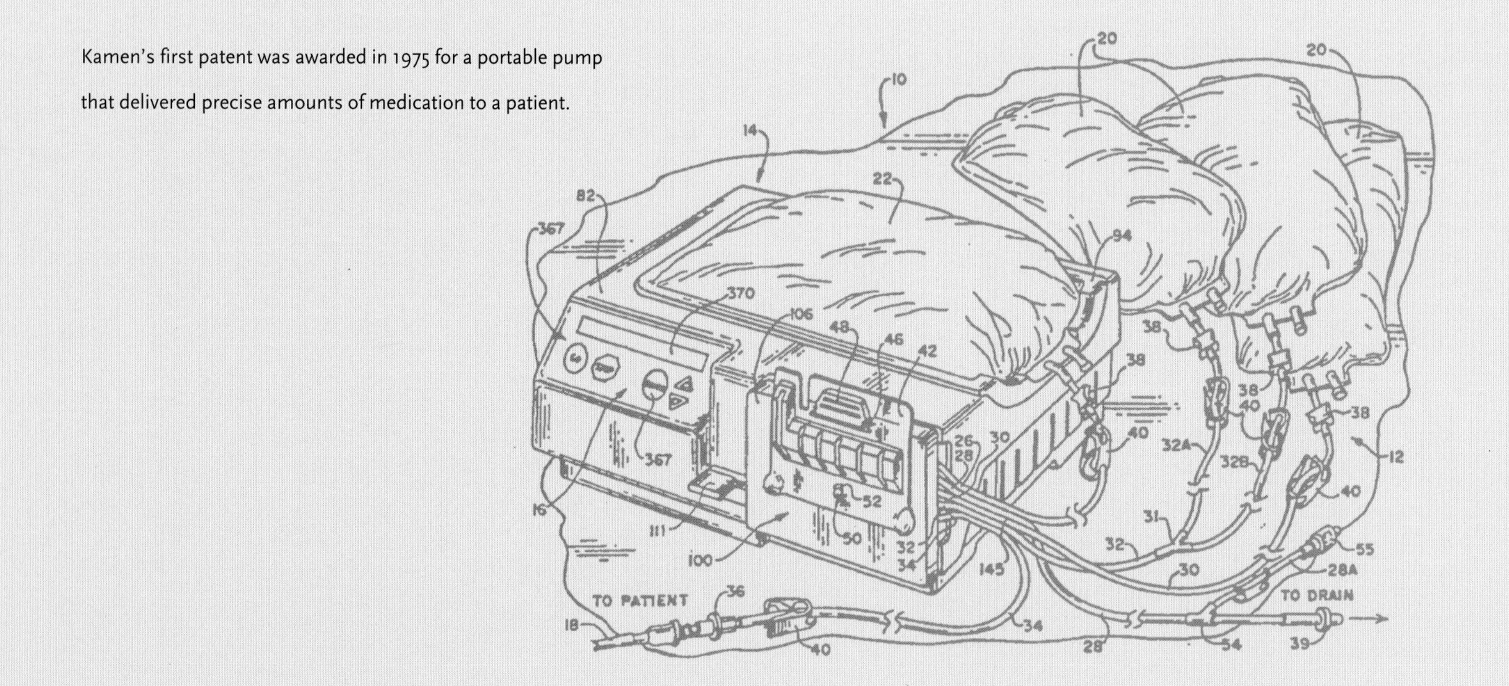

Kamen hangs upside-down in the foyer of his home in order to work on a large, antique engine. His office in Manchester, New Hampshire, includes many other pieces of vintage machinery.

Much of Kamen's attention these days goes to an invention that is not physical at all. In 1989, he started FIRST (For Inspiration and Recognition of Science and Technology), an organization dedicated to developing an interest in technology among high school students. Each year, FIRST partners teams of students with engineers to build robots that compete with each other in regional and national tournaments. The results are auditoriums full of passionate kids shouting encouragement at their robots.

Like many people, Kamen was disturbed by the relative lack of scientific education in American schools. But he didn't focus on the teaching. "I looked at the decay at a cultural level," he explains. "In America you get what you celebrate. Kids will play basketball with chicken-wire hoops because they have desire, they have passion." So FIRST creates a kind of sports for the mind. Its raucous competitions, Kamen continues, "create models and heroes for these kids." A few years ago, Lego joined with FIRST to begin another competition based around its robotics kits. Thousands of students are now involved with the program. All of this pleases Kamen. "The universe is a big playground," he says. "It's fun to be here if you know some of the rules."

MARY-CLAIRE **KING**

Breast cancer strikes more people today than ever before, for reasons ranging from diet to changes in childbearing practices and the use of hormones at menopause. According to the National Cancer Institute, a girl born today in the United States has a one-in-eight chance of eventually being diagnosed with breast cancer if she lives to the age of 95. In 1990, geneticist Mary-Claire King proved that breast cancer can be inherited in some families by mapping the chromosomal location of a gene that leads to about five percent of all cases of the disease—the first time breast cancer had been associated with a single gene. Finding a gene responsible for hereditary breast cancer—and demonstrating how other such genes might be found—holds many promises for women and men with breast cancer. It allows people at high risk to be screened for mutations in the gene; if a mutation is found, the woman who carries it can pursue preventative measures such as frequent mammograms and even preventive surgery. The discovery could also lead to more focused cancer therapies and, ultimately, prevention.

Born in Wilmette, Illinois, in 1946, Mary-Claire King grew up in a traditional family. Her father worked at Standard Oil while her mother took care of the household, including Mary-Claire and her brother. Mary-Claire became interested and skilled in puzzles as a child, an interest that led to a love for mathematics. Because she was so skilled at math, she continued studying it when she enrolled in Carleton College. King knew that studying math could open up all kinds of academic avenues. The question of how to combine mathematics with biological science—an interest sparked by the death of a childhood friend from cancer—was answered by genetics, a discipline in its infancy when she began her doctoral studies at the University of California at Berkeley in 1969.

King had difficulty translating her ideas—such as studying mutations caused by environmental chemicals—into results. Frustrated, she complained to Berkeley biochemist Allan Wilson, a founder of the field of molecular evolution. He told her, "If everyone whose experiments failed stopped

Genetic trails lead to Argentina's missing children

Paula Lavallen was abducted when she was 22 months old. When she was returned to her grandmother's house more than 6 years later, she walked straight to her old bedroom and asked where her doll was.

By Simson L. Garfinkel
Special to the Globe

When Paula Lavallen showed up with her birth certificate and medical documents for her first day of Argentine kindergarten, something was plainly wrong.

Although her parents were not poor, the documents indicated that no medical personnel [illegible] present at her birth. And the birth certificate was signed by a military doctor who had visited her home several hours after Paula's birth.

"The birth certificate was obviously phony," said American geneticist Mary-Claire King, who was called to Argentina to investigate the case. Other [illegible] suspicious, King said; for example, Paula's health had been excellent at birth, even though she had allegedly received no prenatal or postpartum care, and she seemed much older than what her documents reported.

Mayo), a human rights group made up of mothers of young men and women who had "disappeared," along with their small children, between 1976 and 1983, during the rule of Argentina's brutal military dictatorship. In 1984, a government comission documented 8,961 such disappearances.

Many of the disappeared were in their early 20s, said King, a professor of epidemiology at the University of California at Berkeley. Many were women, perhaps a third of them pregnant. Often kidnapped mothers were killed after they gave birth. Other women were kidnapped with infant children in their arms.

Whenever the Grandmothers learned of a child who was suspected of being a kidnap victim – perhaps from an attending midwife, or a [illegible] kidnapp[illegible] put the information into their files and waited. Today, they say they know of 208 children who disappeared during the military regime. Although some of the children were abandoned, they say, many were adopted by military families.

doing science, there wouldn't be any science." Wilson's attitude reassured her, and she soon began doing her doctoral research with him.

This work eventually landed her on the cover of the prestigious journal *Science*. She searched for differences in the genetic code of chimpanzees and humans. Although most people thought there should be significant variation between the species, considering the differences between a living chimp and a human, "I couldn't find any differences," King recalled. There was a difference of only one percent between human and chimpanzee DNA. She and others in the Wilson lab used this data to calculate when chimpanzee and human evolution diverged; the answer was about 5 million years ago, far more recently than in previous estimates.

King began researching the genetics of breast cancer in 1974. The fact that breast cancer ran in some families was recognized as far back as Roman times; the idea was rekindled by clinicians in the nineteenth century. In the 1920s, statisticians began to investigate and write about it. By the 1970s, there was extensive statistical data, but little molecular data, about inherited cancers.

King and her graduate students evaluated the cancer histories of some 1,500 women; about 15 percent of those came from families with a history of breast cancer over several generations. Her team collected blood from families with the highest incidence of breast cancer and used genetic markers to track cancer across generations in each family. A genetic marker is an identifiable stretch of DNA that identifies a specific region on a chromosome. At the time, geneticists had identified only about 30 markers, which limited King's work. In the early 1980s, that number increased to more than 100, allowing the search to be much more specific.

In the 1980s, King applied genetic analysis to a very different subject: identifying the children of *los desaparecidos*, "the disappeared," in Argentina. A group of military generals had taken power in that country after the fall of Isabel Peron's government in 1976. Over the next several years, some 15,000 of their political opponents were kidnapped and killed. Their victims' small children—and the newborn children of kidnapped pregnant women—were often given to childless families of the military or police, or to those involved with the government. Although these children's parents had been killed, their grandparents, especially their grandmothers, began demonstrating for their return in a plaza in front of the government's headquarters in Buenos Aires. The grandmothers also collected information about the children of the disappeared.

In 1983, after the Argentinean military dictatorship was replaced by a democratically elected government, the grandmothers came to America looking for help. They knew where some of the children were, but since all the witnesses to the kidnappings were dead, they didn't know how to prove who was related to whom. Mary-Claire King agreed to tackle the problem. She traveled to Argentina and took blood samples from as many relatives as possible as well as the children suspected of being kidnapped. King compared children's and surviving relatives' blood samples for proteins that are highly variable among individuals—the same proteins that cause humans to reject transplanted organs. With this test, which provided a 99.9 percent level of accuracy, she identified the natural family of Paula Eva Logares, an eight-year-old girl. Since then, some 70 children

In 1990, Mary-Claire King found the general location of a gene responsible for about five percent of all breast cancer. The discovery came after years of laboratory work tracking hundreds of genetic "markers."

THE POLYMERASE CHAIN REACTION

The key to current genetic research is the polymerase chain reaction method (PCR), created in the early 1980s by Kary Mullis. PCR allows scientists to take a small, individual strand of DNA and multiply it millions of times in just a few hours. This abundance of DNA allows researchers to delve more quickly into the mysteries of genetics, lets doctors better diagnose patients, and helps police officers to identify suspects through genetic "fingerprints."

Here's how it works: The fragment or strand of DNA one wants to duplicate is placed in a test tube. When the tube is short segments of DNA—are added. When the mixture cools, the primers attach to the strands in question. Then a large amount of the four individual nucleotides that make up DNA are added to the test tube, along with an enzyme called a DNA polymerase. When the tube is heated again, to 75 degrees Celsius, the polymerase "reads" the DNA sequence and harnesses the individual nucleotides to duplicate it.

When this cycle is finished, there are two new strands of DNA in the test tube, one of each of the original paired strands. As the simple process is repeated, the number of strands in-

have been returned to their relatives; about 150 children are still to be reunited.

Also in the 1980s, genetic technology had gotten a big boost. A new technique, the polymerase chain reaction, allowed researchers to quickly make millions of copies of the DNA they were studying. This enormously increased the number of markers available, from a few hundred to many thousands, making King's search for breast cancer genes much more precise.

King's team laboriously examined these markers one by one, trying to determine if they tracked disease in a cancer-suffering family. In August 1990, they tested their 183rd marker. King found that several families—but not all of them—shared this marker. Looking at some of the data, King thought she might have found the right location for the gene, but wasn't sure.

Then one of her students, Beth Newman, suggested that they include the age at which the people had first been diagnosed with breast cancer, since cancers that occurred at an early age were more likely to be inherited. With this, King said, "everything fell into place." The marker she had found tracked disease in families with the youngest patients; after that, the correspondence was weaker. Women with the still hypothetical, but now mapped, gene had an 85 percent chance of developing breast cancer over their lifetime; to them, it was potentially lifesaving news.

King announced her findings at a genetics conference in October 1990. The news sent shock waves throughout the research community. Immediately, other scientists rushed to verify her research. There was still a huge amount of work to do: King had identified only the marker for the breast cancer gene (which she named BRCA1), not the gene itself. In the marked stretch of chromosome, there were some 50 million pieces of genetic code—pairs of the bases that make up DNA. The race to find the exact gene had begun; with King's research pointing the way, anyone could find it.

One of the first results of this new wave of study was the discovery by the French geneticist Gilbert Lenoir that the same markers tracked ovarian cancer in some families. Then, in the summer of 1994, a team of scientists that included Roger Wiseman, Andy Futreal, and Mark Skolnick found BRCA1. Since then, another breast cancer gene (BRCA2) has been found, and research continues on other genetic causes of cancer.

The discovery of BRCA1 and BRCA2 has led to genetic screening for people from families with a history of breast cancer. For many people, it is important to know how high their real risk for cancer is. Those who do harbor mutations in one of these genes know to watch much more closely for signs of cancer; some have chosen to undergo phrophylactic surgery. The isolation of the breast cancer gene also holds the possibility for gene-targeted therapy, where an outside agent enters the body's cells to alter the consequences of losing the critical gene. Although effective therapy based on these genes against a disease as complicated as breast or ovarian cancer is still years away, it is King's work that has opened the door.

creases exponentially, doubling each time. In about three hours, the test tube will contain millions of copies of the DNA.

With the creation of PCR, scientists could make endless, perfect copies of any bit of DNA they were interested in. Suddenly, the limitations of time (it took a long time to isolate a bit of DNA) and supply were gone. PCR has not only allowed larger scale genetic investigations, like the Human Genome Project, but the simple method has also brought the cost of DNA analysis way down, allowing for much more precise medical testing in labs all over the world.

ROBERT **LANGER**

When chemical engineer Robert Langer began working at Children's Hospital in Boston in 1974, the synthetic materials being used for medical procedures were fairly primitive. If doctors needed to fashion an artificial heart valve, for example, they searched for an off-the-shelf material that had the elasticity of the actual heart. They found it in a ladies' girdle.

Over the past 25 years, Langer has introduced several new medical materials that not only are far more sophisticated than a girdle's polyurethane, they also work much better. Langer's most notable advances have come in the field of drug delivery, where he created several polymers and other systems that can administer precise doses of drugs to the right part of the body. His polymer research also led him to tissue engineering, the manufacture of replacement skin, cartilage, and bone. Today, Langer and other scientists are working to grow whole organs in the lab, from the bladder to the liver.

Born in 1948, Robert Langer grew up in Albany, New York. His interest in chemistry was sparked when he received a Gilbert chemistry set as a teenager. "I just was fascinated playing with [the chemistry set]," he remembers, "seeing the colors change when I mixed different things together." Since he was good at math and science, Langer was encouraged to study engineering, and he enrolled at Cornell University. The only subject that really interested him, though, was chemistry, so he concentrated on chemical engineering. He earned his bachelor's degree in 1970, then a doctorate at Massachusetts Institute of Technology in 1974.

Langer received about 20 job offers when he finished his graduate work, but none of them really appealed to him. When he was offered a postdoctoral fellowship with Dr. Judah

Michael Cima, John Santini, and Robert Langer display a silicon wafer holding several of their programmable drug-delivery chips. Each chip contains dozens of tiny reservoirs for drugs; once implanted, the device can release precise amounts of medicine.

Folkman, who was doing cancer research at Boston Children's Hospital, he jumped at the chance. It was an odd position for a chemical engineer; while today it is not uncommon for engineers to work in medical research, it was almost unheard of at that time. "People thought I was nuts, and they probably thought Judah Folkman was nuts, too," Langer says.

Dr. Folkman's lab proved a fertile environment for Langer. "I started getting a lot of ideas about how you could apply chemical engineering to medical problems," he recalls. One of the areas he began exploring was drug delivery. At the time, few people were exploring the science of gradual or controlled delivery of drugs into the body. Pharmaceutical companies had figured out how to time-release small-molecule drugs by encasing them in wax that broke down over time; the best known example of this is the Contac cold capsule, which contains hundreds of encapsulated "pellets" with its active ingredients.

Langer thought that synthetic polymers, which he had worked with in school, could hold a new solution to delivering drugs—especially those, such as steroids and hormones, with larger molecules. Although synthetic materials had already revolutionized plastics, fabrics, and other industries, there was little or no original work being done in creating polymers expressly for medical applications. In addition to the girdle material used in heart valves, Langer notes that one type of breast implant was made from an industrial lubricant while another came from mattress stuffing.

The engineer's first invention to come out of this investigation, conducted with Dr. Folkman, was a polymer structure that could slowly release almost any kind of drug molecule. "We were able to create a very complex, tortuous, porous network in the polymer," Langer says. Drug molecules "wanted" to get out of the polymer, but the difficult internal structure made their progress slow. He compares the process to driving in Boston, a notoriously difficult activity. Since one could measure the drug molecule's progress, the polymer acted as an accurate and gradual way to release drugs into a patient.

After the success of this complex polymer, Langer knew that he could create custom materials with almost any kind of molecular properties. "I thought we could come up with a strategy where we could ask what you really want from a biomaterial and then design it from scratch, take it from first prin-

Since 1974, Robert Langer has been applying his chemical engineering skills to medical problems, especially the challenge of delivering drugs in precise doses over a period of time.

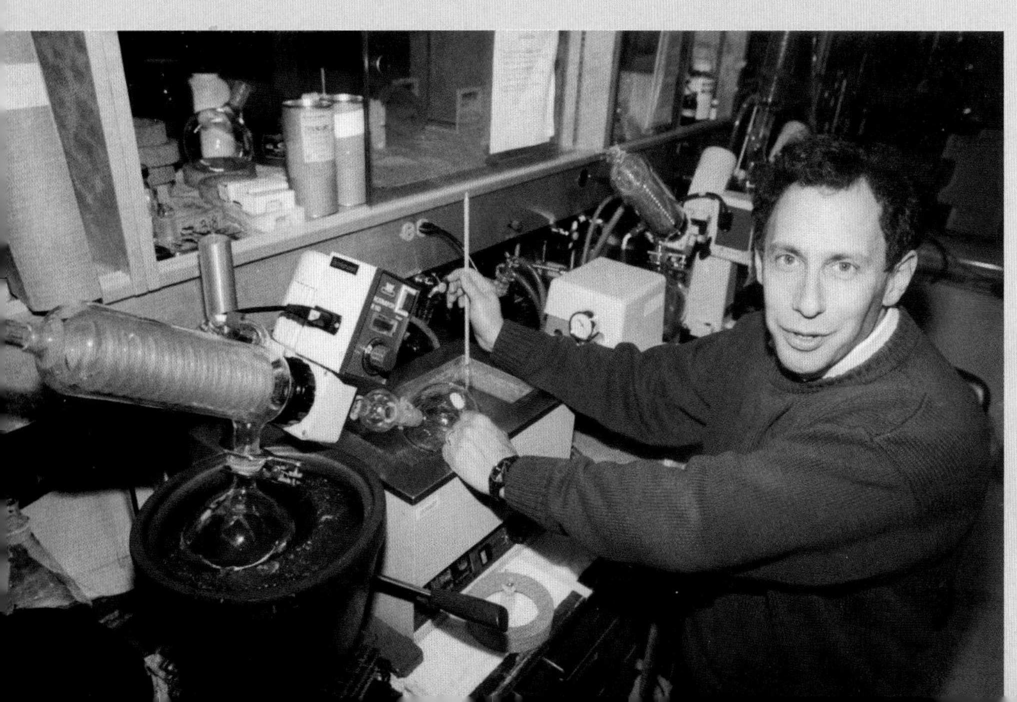

ciples," he says. Langer and his students at MIT, where he has taught since 1977, decided to try to make a polymer that would degrade over time—a property that would gradually release encased drugs. They succeeded with a type of polymer called a polyanhydride.

When Langer first made polyanhydrides, he wasn't sure what their specific applications might be. Then Dr. Henry Brem of Johns Hopkins University approached him with an idea for using the new polymer to treat brain cancer. Chemotherapy for brain cancer was a dangerous process. The human body protects the brain with the so-called blood-brain barrier, which keeps most drugs and other harmful substances out of it. The traditional way to get around the barrier is to load the body with large amounts of drugs, a small portion of which will reach the brain. Cancer drugs, however, are highly toxic, and this method could damage the kidney, liver, and spleen.

Langer and Brem devised a drug-delivery system that not only minimized the amount of drugs in the body, but also concentrated them directly at the point of disease. They created a small wafer of polyanhydrides and BCNU, the drug used to fight brain cancer. A brain surgeon would remove as much of a cancerous tumor as possible, then implant up to eight of the dime-sized wafers into the brain. The polyanhydride biodegrades over the course of a month, slowly releasing the drug right at the affected area of the brain.

When the Food and Drug Administration approved Brem's and Langer's drug wafer in 1996, it was the first new treatment for brain cancer in 23 years. One recent study of its effectiveness showed that five times more patients survived over two years than with conventional treatment.

Biodegradable polymers also led Langer into a different area of research: tissue engineering. He and Dr. Joseph Vacanti of Harvard University began working on growing human skin in the lab in the 1980s. Skin cells were grown on a two-dimensional frame made of polyanhydrides. As the cells multiplied and became a coherent tissue, the polymer frame biodegraded, leaving only the tissue. Working in two dimensions was limiting—they were able to grow only tissues about 1/16th of an inch thick. When they tried to make them thicker, the tissue on the inside couldn't get enough oxygen or nutrients.

In 1986, Dr. Vacanti was watching seaweed wave in the water off Cape Cod when inspiration struck: If they could create a polymer frame that mimicked the branching structure of the seaweed, they might be able to create three-dimensional tissues. Langer made just such a framework, and it did work. One researcher has already grown bladder tissue using the technique, and Langer and others have created working liver tissue. Someday, Langer says, they may be able to create whole, functioning organs for use in humans.

Recently, Langer has been working on an even more sophisticated drug-delivery system. While watching a documentary on how computer chips are made, he realized he might use a similar technique to precisely deliver drugs. Working with computer engineer and MIT professor Michael Cima and then graduate student John Santini, he designed a device that could both store and release minute quantities of medicines. On a chip the size of a dime, there are 34 small reservoirs for drugs, sealed on both sides with gold plates. A microprocessor can send an electrical impulse to an individual reservoir. The interaction of the gold, electricity, and a chloride solution around the chip causes the gold plate to corrode, releasing the drug. The chip can be programmed to open the reservoirs in any order, at any time, allowing for precise and complex drug-delivery protocols. Although the prototype has just 34 reservoirs, the chip at its current size could eventually contain 1,000 microdoses of drugs, even more if the chip was bigger or the reservoirs smaller.

The next challenge Langer is facing is gene therapy, the introduction of DNA into the human body to combat various diseases. Until now, most attempts have used viruses to deliver the DNA to the body's cells. There are obvious dangers in using viruses as a delivery system; in 1999, one patient died during a trial of this kind of gene therapy. Langer's method concentrates on benign polymers. He and his collaborators are trying to create a polymer small enough to move through the wall of a cell and that also has chemical characteristics that will make it stop at a certain part of the cell, where it will release a payload of DNA.

Although the fields Langer has helped create—biomedical polymers and tissue engineering—are barely 25 years old, they have already benefited millions of people. As more and more engineers, doctors, and researchers join the fields, how medicine works may be fundamentally changed. All of this helps Langer pursue his original goal: "I was always interested in doing things that I thought would help people."

ROSALYN YALOW

Rosalyn Yalow's path to her 1977 Nobel Prize in Medicine was not an easy one. As a woman, she was discouraged or prevented from pursuing her education. And even when she and her longtime collaborator, Dr. Solomon Berson, made a huge breakthrough in studying diabetes, scientific publications wouldn't print their conclusions, considering them too revolutionary. It was that breakthrough, though, that led to their invention of radioimmunoassay (RIA), an extremely sensitive method for measuring substances in the body. RIA led to major medical discoveries and is now used in research and medical labs all over the world.

Rosalyn Yalow was born Rosalyn Sussman in 1921 in New York City, where she has remained for all but a few years of her life. Her father ran a small business on the Lower East Side of Manhattan selling paper and twine; during the depression, her mother did sewing work at home to make ends meet. Although neither of her parents had a high school education, they always encouraged her academic pursuits. Yalow's love for knowledge was clear very early; she began reading before kindergarten, and she and her brother would walk to the library once a week to trade in their just-read books for fresh ones. Her legendary stubbornness was also evident from a young age. "Perhaps the earliest memories I have are of being a stubborn, determined child," she wrote. "My mother has told me that it was fortunate I chose to do acceptable things, for if I had chosen otherwise no one could have deflected me from my path."

When Yalow began attending Hunter College, New York's public women's college, she fell in love with physics. "In the late 1930s," she wrote, "physics, and in particular nuclear physics, was the most interesting field in the world. It seemed as if every experiment brought a Nobel Prize." Yalow was such a driven student that Hunter created a physics program just for her.

When she graduated from college in 1941, Yalow's options were limited—very few graduate programs in physics would admit women. As she looked for some way to continue her education, she was offered a part-time job as a secretary for a biochemist at Columbia University; she would be able to take some graduate courses there, as long as she learned to take shorthand. She

Yalow's breakthrough—the development of radioimmunoassay, a highly sensitive technique for measuring substances in the body—came in the late 1950s. She continued working as a researcher, and later as a professor, until her retirement in 1991.

Rosalyn Yalow built the Bronx Veterans Administration Hospital's nuclear medicine research program almost from scratch. Here she prepares an "atomic cocktail" of radioactive isotopes in 1948, her second year at the hospital.

soon enrolled in secretarial school. But the country was preparing for World War II, and suddenly male graduate students were in short supply. Yalow was finally offered a place at the University of Illinois, and she tore up her stenography books.

Things were not much easier in Illinois. Yalow was shocked to discover that she was the only woman in a class of 400. (On the brighter side, on the first day of class she met fellow physics student Aaron Yalow, whom she would marry in 1943.) And Yalow's course work in physics at Hunter had been minimal—less than any of the other graduate students. She set to work, taking undergraduate courses to catch up and graduate courses, while also teaching physics to freshmen. She received three As and an A- in an optics laboratory; the chairman of the department declared, "that A- confirms that women do not do well at laboratory work."

Yalow earned her doctorate in 1945 and returned to New York. While she was teaching physics to returning veterans at Hunter College, Aaron Yalow began working as a radiation physicist. Soon, he introduced Rosalyn to Dr. Edith Quimby at Columbia's medical school, who was researching the medical applications of radioactive isotopes—artificially made, unstable variants of elements like carbon and iodine. Quimby had just the job for her: She called up a colleague at the Veterans Administration hospital in the Bronx, and Yalow was soon working in its brand-new nuclear medicine program.

The era when Yalow began working at the V.A. was one of high hopes for nuclear physics. Although most people were horrified by the atomic bombs dropped on Hiroshima and Nagasaki, some worked to apply nuclear physics in technologies that would benefit mankind. President Eisenhower announced the "Atoms for Peace" program, part of which was Project Plowshare, an attempt to find peaceful uses for nuclear bombs. Weapons scientists proposed using bombs to dig harbors, release underground deposits of fossil fuel, and create a new canal through Panama, among other things. (The dangers of radioactive fallout kept any of these from happening.) On the medical side, many researchers were exploring the uses of radioactive isotopes to treat cancer and other diseases.

Yalow built the V.A.'s radioisotope research program almost from scratch. Working out of what had been a janitor's closet, she built her own equipment and began clinical research. By 1950, she knew she needed a partner with medical knowledge to complement her work on the physics side. That partner was Solomon Berson, a young doctor whom Yalow later called "the smartest person I had ever met." Their close and productive professional relationship would last until his death in 1972.

Yalow and Berson began to use radioisotopes to measure the hidden workings of the human body. This led them to look closer at the endocrine system, which is composed of the endocrine glands (the thyroid, adrenals, ovaries, testes, and others) and the hormones they produce. The endocrine disorder that affects the most people is diabetes. Diabetics are not able correctly to process insulin, a hormone that keeps a person's blood sugar in check. No one knew exactly what happened to insulin in a diabetic; Yalow and Berson spent hours discussing the problem in their office, where they had positioned their desks head-to-head so they could speak directly to each other.

The two researchers thought that radioactive isotopes might help reveal how insulin (and thus diabetes) worked. By tagging an individual substance, in this case insulin, with a radioactive isotope of iodine, they could track the otherwise invisible hormone by measuring its radioactivity. When they injected tagged insulin into diabetics, they observed something unexpected: Their insulin became bound with a larger molecule, gamma globulin. Yalow and Berson determined that the gamma globulin was an antibody, a protein the body's immune system produces to protect us from invaders such as viruses and bacteria. When a diabetic was injected with synthetic insulin, they discovered, the body was treating it as an invader, preventing it from doing its usual job. This groundbreaking study was published in 1956, and it led to the development of synthetic insulin that is not attacked by the immune system.

Even more significant than what they found was how they found it. Buried in the middle of their paper on insulin was a description of their new method for measuring the insulin, a method that they eventually called radioimmunoassay (RIA). The basic principle of RIA works for a number of substances, from hormones like insulin to enzymes, vitamins, viruses, and drugs. First, a researcher tags a known sample of a hormone or other substance with a radioactive isotope. Then, that hormone is mixed with its naturally occurring antibody to form a solution of hormone-antibody pairs. This solution is combined with a quantity of a patient's blood, which contains an unknown quantity of the hormone being tested for. Antibodies "prefer" nonradioactive versions of the hormone they were made to match up with, so after a period of time they let go of the tagged hormones and pair up with the natural ones in the patient's blood. A researcher then separates out all the hormone-antibody pairs (both radioactive and not) from the solution. By calculating the difference in radioactivity of this group of pairs with the original tagged group, one can measure how many radioactive hormones were replaced by natural ones—which is to say, how many hormones were originally in the patient's blood.

This may seem like a complicated procedure for a simple result, but before RIA it was essentially impossible to measure substances as tiny as insulin. And RIA is extremely sensitive; it can measure very low concentrations of these substances, which is important because they generally occur in the body in very small quantities. Insights from RIA testing have led to new treatments for diabetes, mental retardation, and other conditions.

In the 40 years since its invention, RIA has become one of the basic methods of medical testing and research and has

WOMEN IN SCIENCE

Rosalyn Yalow wasn't the only woman scientist who found opportunity during World War II. Graduate schools, university faculties, and military and civilian research labs suffered a severe shortage of able-bodied and able-minded scientists as much of the young male population of the country enlisted or was drafted into the armed services. And, just as industrial jobs were famously filled by "Rosie the Riveter," women quickly replaced men in scientific fields. The war was one of the first cracks in the all-male—or at least almost all-male—surface of science. For many women, however, wartime gains were temporary. "When the men returned to civilian life," notes historian Martha Bailey, "the women were expected to return home."

While the end of the war curtailed many women scientists' budding careers, it didn't end all of them—four of the 11 women who have ever been awarded a Nobel Prize for science received some career boost from World War II. And the war also had profound effects on science in general—effects that would help open its ranks to a much more diverse group. Many disciplines and technologies were introduced or matured then, including nuclear physics, advanced medicine, and computing, and these fields required a large number of new workers, too many to be filled only by men. Another barrier the war helped demolish was one of cultural expectations: Most people assumed that scientists were men, often lone geniuses. As thousands of women and men entered scientific and technical professions for the first time, science was demystified. "A profession that had been cohesive, homogeneous, and populated in the hundreds," writes Vivian Gornick in *Women in Science*, "now exploded with uncontrollable multiplicity into a pluralism that soon numbered in the hundreds of thousands."

Even though the stage was set for science to become diverse, it would take decades before it really happened. Throughout the 1950s and 1960s, many women were thwarted from entering scientific careers, some by outright discrimination, some by the effects of the "old-boy" network, and some by ingrained attitudes toward women's roles in society. (Many women were asked if

been adopted by virtually every medical lab in the world. Variations on the basic method have also been developed to measure an even broader range of things: The standard test for HIV, for example, is done with a nonradioactive version of Yalow's and Berson's groundbreaking invention. And while the two researchers knew in 1956 that radioimmunoassay was an important medical advance, they never considered making money from their invention. "We never thought of patenting RIA," Yalow told Eugene Straus, her friend and biographer. "Patents are about keeping things away from people. . . . We wanted others to be able to use RIA."

they intended to marry, Bailey notes, or when they were going to start a family, implying their employment might be temporary.) Antinepotism rules kept wives and husbands from working at the same university, usually dooming the wife to a temporary position, a job at a lesser institution, or no job at all.

Progress in bringing women into science careers speeded up in the late 1960s and 1970s, as a wide range of social attitudes shifted and the generation of women born after World War II—women who had internalized new definitions of what science is, and what work should be—came of age. Efforts to achieve equality in the workplace (and throughout society) slowly paid off, as more and more women became scientists or pursued other professions. Between 1966 and 1996, the proportion of women earning bachelor's degrees in science or engineering jumped from 24.8 percent of the total to 47.1 percent. The number of female science and engineering Ph.D.s saw an even greater jump: from 8 percent to 31.8 percent.

While many women going into science during the twentieth century aimed toward university jobs, this boom in science education has led to more diverse goals and opportunities, notes Catherine Didion, executive director of the Association for Women in Science. "Today, a lot of technical and scientific women are moving into entrepreneurial and small businesses, such as smaller firms supporting biotech, engineering consulting firms, places where you need a degree but one has a lot more control over one's day and how it's spent. Women are seeing those degrees as a tool for more than just academic careers, which wasn't the case even two decades ago."

CONSUMER PRODUCTS

INTRODUCTION

In the late nineteenth century German engineer Wilhelm Maybach, at the time working with the Daimler car company, put together the new perfume pump spray with the new gasoline and came up with what we now call the carburetor.

Maybach is a good example of what may be the most fundamental of all elements in the magic process of invention: the moment when two things are brought together, by an innovative mind, in a new way that causes one and one to equal three (the result being more than the sum of the parts). Sometimes, as in the case of one of our inventors in this group, the different parts of the equation are quite bizarre: Percy Spencer's microwave oven owes its existence, as much as anything, perhaps, to the coming together of a magnetron and a chocolate bar.

The best-laid plans often go awry and generate innovation. In 1852 British chemist William Perkin went looking for artificial quinine in coal tar. After weeks of noodling he came up with a dark sludge that was definitely not quinine. On throwing it into the sink, Perkin discovered he had unintentionally invented the world's first artificial aniline dye.

The same kind of off-center result came out of the work of Stephanie Kwolek, when she decided to grapple with an impending gasoline shortage by finding a way to make tires lighter and stronger (to save weight and fuel). As you'll see, that wasn't all she found.

Daydreams also do the trick. The great German chemist Kekulé von Stradonitz said he conceived of the benzene ring while dozing on a London bus, where he dreamed about serpents eating their own tails. There's no suggestion here that Philo T. Farnsworth was asleep on the job, but the image of a hayfield at harvest time seems to have figured large in his invention of the line-scan system fundamental to every television set.

Sometimes, a basic principle will find innovative application in many different fields and circumstances. The seventeenth-century discovery of the vacuum created just such a situation. It gave rise, among many other things, to ventilators for hospitals, the barometer and weather forecasting, Galvani's research into electricity (leading Volta to the battery), and Newcomen's pre-Watt steam engine. Taking a principle, and seeing what it can be made to do, is one technique adopted here by Jerome Lemelson, who proves his talents in the matter with a list of over 500 patents.

Curiosity is a powerful driver of inventive thinking. In 1304 the medieval monk Dietrich of Freiberg virtually started the modern study of optics with his investigations into why the rainbow looked as it did. His sheer curiosity (there was no possible practical application for the research at the time) led him to shine light through simulated raindrops made of small flasks filled with water and to come close to understanding the mechanisms involved.

The feeling that inventive opportunities lie out there waiting to be identified was well expressed by one of our inventors, Jacob Rabinow: "I was willing to bet lunches for years that *something* inside the Post Office could be automated." As a result of that hunch, your mail is now sorted by machines.

Jerome Lemelson believed that inventors are powered by an insatiable curiosity and that they best satisfy this by working alone. In some way or other, however, all the inventors in this group have come to experience what can be the stultifying effect of corporate or bureaucratic institutions. As has often been said, a camel is a horse designed by committee. This problem has existed since prehistoric stone weapons and tools generated the first top-down hunt-planning hierarchy and the "standard operating procedure" that has so often since stifled the innovative spark.

Ironically, institutionalized thinking is all too often the result of innovative thinking in the first place. The first clay-tablet writing in early Mesopotamia made possible a level of organization that immediately required laws. In next to no time this produced a legal system every bit as cumbersome as the one with which we suffer today. The institutional problem arises in the first place because when an innovation makes life better for both the individual and the group, which is the aim of every innovator, one reaction is frequently to confine the new idea or the process, so as to protect it from any interference that might limit its beneficial effects. All too soon those responsible for this happy state of affairs become more interested in protecting the idea from any input at all, fearing it might rock the boat. This in turn often generates an attitude expressed as: "This is how we do things here," which can render further innovative thinking impossible.

Jerome Lemelson also once said that the work of the inventor-entrepreneur is essential to the well-being of the United States. Innovation generates jobs. In the 1930s the U.S. Department of Labor identified 80 different types of work (education, medicine, engineering, secretarial, and so on). Today, seven decades of innovation later, the number of types of work has risen to over 800. The U.S. marketplace has exploded beyond anything that could have been imagined in the 1930s, and we stand at the threshold of an e-commerce global consumer goods revolution that will dwarf all that went before. The requirement to satisfy that market with innovations will be well-nigh insatiable. Minds like those of this group of inventors will have much to occupy themselves with for the foreseeable future.

—JB

LEO BAEKELAND

At the beginning of the twentieth century, there was no such thing as plastic. The word itself was an adjective, not a noun, describing the quality of being moldable or changeable. Bakelite, the material Leo Baekeland created in 1909, gave it a new definition. Capable of being molded into almost any shape, Bakelite was indeed "plastic," but when the multitude of products made out of it found a permanent place in people's homes, the material's basic quality was transformed into its very identity. More important, the material and its descendents eventually transformed the world, making many products, and their inexpensive manufacture, possible, from Saran Wrap and Tupperware to space suits and portable electronics.

Born in Ghent, Belgium, in 1863, Leo Baekeland was the son of an illiterate shoemaker who hoped the boy would continue his trade. The young Baekeland, however, had broader interests, notably chemistry, and he excelled at his studies. At the age of 16 he won a scholarship to the University of Ghent, and by the time he was 21 he had a doctorate in electrochemistry. A few years later he became a professor of physics there.

Baekeland was also an avid photographer, and in the 1880s he began looking at ways to improve the photographic process. At that time, photographs were made on "dry plates"; pieces of glass coated with a light-sensitive emulsion were exposed, then immersed in a chemical "developer" in a darkroom. Baekeland invented a dry plate that had the developing chemicals built-in—when immersed in water, the plate would develop itself. For this he received his first patents, and in 1888 he founded a company to manufacture the plates. The business soon interfered with Baekeland's academic career, and he was faced with a choice between the two—a decision complicated by his engagement to the daughter of the head of his physics department.

In 1889, Baekeland and his bride, Celine, sailed to the United States on their honeymoon. While in New York, he met with the owners of a photographic company and decided to work for them, abandoning both teaching and Belgium. In the 1890s, he began working on a new

The key to manufacturing Bakelite was a special vat, which Baekeland dubbed the Bakelizer. Half boiler and half pressure cooker, the device allowed Baekeland to monitor and adjust the pressure and temperature of the process.

BAKELITE AND MODERNISM

Leo Baekeland intended for his plastic to change what things were made of. But as with many technologies, Bakelite had an unintended consequence: It also changed what things looked like. The almost infinitely moldable plastic was perfect for the aesthetic styles—Art Deco, Streamline, Modernism—just being introduced.

The basic properties of Bakelite encouraged a streamlined look. Until moldable plastic came on the scene, many products were made by joining pieces of wood or metal or other materials—so things had right angles and seams. By contrast, to make something out of Bakelite one poured the resinous plastic into a mold. To ensure that it flowed smoothly, designers had to avoid sharp angles and edges in their molds. The results were products with smooth, gentle curves.

Bakelite also became popular just as a new discipline was emerging: the professional designer. Previously, product design had been the job of engineers, or craftsmen, or no one at all. This changed in the late 1920s and 1930s, as schools such as the Bauhaus began to teach design. One of the beliefs of this first generation of modern designers was that "form follows function"—that a product's design should reflect what it does. Fancy materials and decorations were eliminated, while shapes were simplified. And in their rejection of previous styles, modern designers embraced the "machine age"—they celebrated all things man-made, especially synthetic materials.

One of Bakelite's most far-reaching impacts is easy to overlook. When first introduced, it was seen as a cheap imitation of "real" materials like wood—in fact, many early Bakelite products came in mottled brown, to imitate wood. As ever more elaborate and colorful goods were made with it, though, Bakelite became a "real" material itself. Designers were finally free to create purely synthetic goods, and they never looked back.

Advertisements called Bakelite the "Material of a Thousand Uses." Today, Bakelite is best known for its role in consumer products of the 1930s, especially brightly colored plastic housewares and jewelry.

kind of photographic paper, and through careful experiments and much trial and error, he invented one that was much slower than previous papers—so slow that photographers were able to use it under artificial light in a darkroom, where they had much more control over the results. Baekeland's paper, which he named Velox, appealed to the new market of amateur photographers and achieved great popularity. The Eastman Kodak Company, already a photography giant, saw it as competition and purchased the rights to the paper for $750,000—a huge sum for the time.

One requirement of Baekeland's agreement with Eastman Kodak was that he not work on anything photographic. With that part of his career cut short and a great deal of money in the bank, Baekeland bought a large estate in Yonkers, New York, where he set up his own research laboratory. There he turned his attention to a problem other chemists were attacking: inventing an artificial substitute for shellac.

Shellac, a resin harvested from a Southeast Asian beetle, was used as an electrical insulator, among other things. As America began to use electric power on a large scale, demand for shellac outstripped the supply, and prices went up. The race was on to find a substitute, a material that wouldn't conduct electricity and was plastic—easy to shape.

People had been pursuing such a material for years. In the eighteenth and nineteenth centuries, powdered horns and hooves were molded into snuff boxes and other goods. In the 1840s, the vulcanization process turned natural rubber into a more durable and moldable material. And in the 1860s, as the world faced an ivory shortage owing in part to a craze for billiards, the American John Wesley Hyatt created celluloid, made from plant fiber and camphor. It could be permanently molded into any shape, and it found a market as a substitute for ivory, bone, and other materials. But celluloid was flammable—billiard balls made from it would occasionally explode—and perceived as a cheap substitute for better materials.

Baekeland focused his research on finding a material that was purely synthetic—using no natural materials. In 1904, he and his assistant, Nathaniel Turlow, began investigating the products of combining two chemicals, formaldehyde and phenol, a product of coal tar. Thirty years earlier, a German chemist had worked with the two chemicals in pursuit of a synthetic dye, but the odd and unmanageable resin they produced was not of interest to him. By the turn of the century, other researchers had combined formaldehyde with casein, a substance derived from milk, to make an early plastic, but its use was limited.

Baekeland methodically studied each variable that went into the reaction between phenol and formaldehyde—from the proportion of chemicals to heat and pressure, as well as the effect of added solvents. The process usually created a useless goo, sometimes too brittle and sometimes too spongy. Baekeland's breakthrough came in 1907, when he learned to control the gases given off by the chemical reaction. In a complicated apparatus that was half boiler, half pressure cooker (it was later dubbed the Bakelizer), Baekeland sped up the reaction with heat and kept the gases in check with pressure.

It took several steps to turn the translucent solution that came out of the Bakelizer into Bakelite. First, it was allowed to cool and harden, then it was ground into a powder. A filler composed of wood powder, cotton, asbestos, or other materials was added, as were other chemicals necessary to the next stage. When heated, the mixture turned into a resin that was again allowed to cool, then ground up. This powder was Bakelite, ready to be put into a mold, then set in place forever with the application of heat and pressure. The final product was "a magnificent and hard mass," in Baekeland's words, impervious to heat, electricity, and most solvents. In his lab book, he wrote, "Unless I am very much mistaken this invention will prove important in the future." The chemist soon filed two patents covering the process, the first of many protecting his invention.

Almost two more years of research and refinement passed before Baekeland was ready to announce his invention. In February 1909, he presented Bakelite to the New York chapter of the American Chemical Society, showing off several items made from it. The assembled chemists gave him an ovation, and the era of plastics was underway. At first, Baekeland envisioned Bakelite as an industrial material, to be used for electrical insulators, floor tiles, and chemical- and fire-resistant coatings, among other things. Its "use for such fancy articles [knobs, buttons, handles]," he told the American Chemical Society, "has not much appealed to my efforts as long as there are so many more important applications for engineering purposes."

The electrical industry eagerly adopted Bakelite, using it for sockets, fuses, switches, and other small molded components. By the early 1920s, the new radio and automobile industries we using it, too. The material was flexible, durable, and colorful

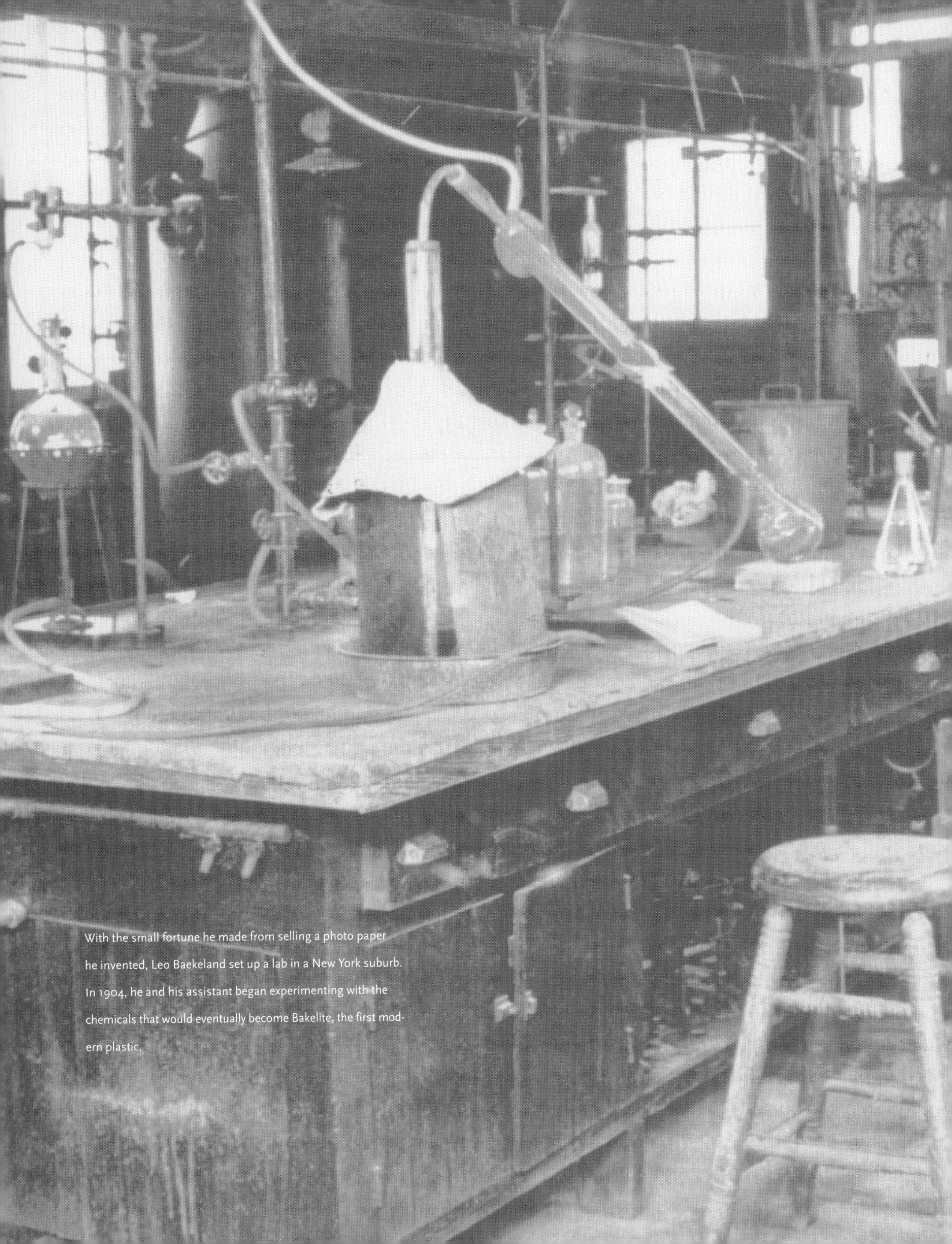

With the small fortune he made from selling a photo paper he invented, Leo Baekeland set up a lab in a New York suburb. In 1904, he and his assistant began experimenting with the chemicals that would eventually become Bakelite, the first modern plastic.

(dyes could be added to produce vibrant hues such as yellow, orange, and red), and it suggested progress to the consumer.

Meanwhile, Baekeland had entered a long patent war with other chemical firms, which had started to manufacture Bakelite knock-offs almost as soon as the invention was announced. One strategy of Baekeland's company—the General Bakelite Corporation—was to brand products with a logo. A more successful approach was taking competitors to court; after a long battle, Baekeland orchestrated a merger of plastics manufacturers. In 1939 he sold the company to Union Carbide and retired to Florida. He died in 1945.

As the 1920s went on, an increasing number of products were made out of Bakelite. Radios and telephones were the most visible of them, but the material soon entered almost every consumer niche. Indeed, the product's slogan was "The Material of a Thousand Uses." There was Bakelite jewelry, kitchenware, cigarette cases, cabinets, lamps, toothbrushes, and fountain pens. Even after America plunged into the depression, sales of Bakelite products remained strong. Not only were they colorful, durable, and attractive, but mass production made them inexpensive.

Bakelite remained popular until World War II, when production lines were switched over to war materials, not consumer goods. It was also losing its technological edge. In the 1930s, DuPont had introduced its first synthetics, including Lucite, Plexiglas, and Nylon; after the war, these and other new plastics took over the market.

Today, of course, plastics define the look and feel of almost everything we touch. And a new wave of plastics, including some that (surprisingly) conduct electricity, will make synthetics an even more fundamental part of the material world. With Bakelite now known more as a collectible than a material that changed the world, it's easy to forget that the plastics revolution began with Leo Baekeland's research in a vat called the Bakelizer.

HAROLD "DOC" EDGERTON

Harold Edgerton's business was stopping time. With high-speed cameras and flash equipment of his own invention, Edgerton photographed the events of a split second—a thousandth of a second, even a millionth of a second. By capturing those tiny moments of time he revealed, often for the first time, how things work. In his pictures, a hummingbird's wings stop; a drop of milk becomes a beautiful, crown-shaped splash; a golf ball deforms on impact; and a nuclear explosion expands, microsecond by microsecond, into a huge fireball.

Edgerton's photographs and equipment were important to science and business, revealing in sharp detail the smallest increments of industrial and natural processes. But his images were even more appealing to the public, which was thrilled to see the hidden world they unveiled. His pictures were first published in *Life* magazine and elsewhere in the mid-1930s, a time of great technological change and hope. Atomic science was beginning to explore the innermost workings of the universe, while new electronics, drugs, and other technologies were improving the quality of Americans' lives. Edgerton's photographs were a kind of proof that this new world was understandable, that science and technology could help us comprehend previously mysterious events. They were also often beautiful; many of his photographs are now in major museum collections.

The man who opened these hidden realms to public view had humble beginnings. Edgerton, who was known to most people as "Doc," was born in 1903 in Fremont, a small Nebraska farming town. His father held a succession of jobs, including president of a local bank, school principal, lawyer, and Washington, D.C., correspondent for a newspaper in Lincoln, Nebraska. Eventually, the family settled down in Aurora, Nebraska, population 3,000. The young Edgerton became interested in all kinds of mechanical things, from cars and motorcycles to engines and electrical equipment. When he was in high school, he went to Aurora's electrical plant and asked for a job. The future inventor was hired to sweep the floors. His mechanical talents soon

With the electronic flash he developed for industrial use, Harold Edgerton could capture even the briefest of moments.

Hummingbird, late 1940s

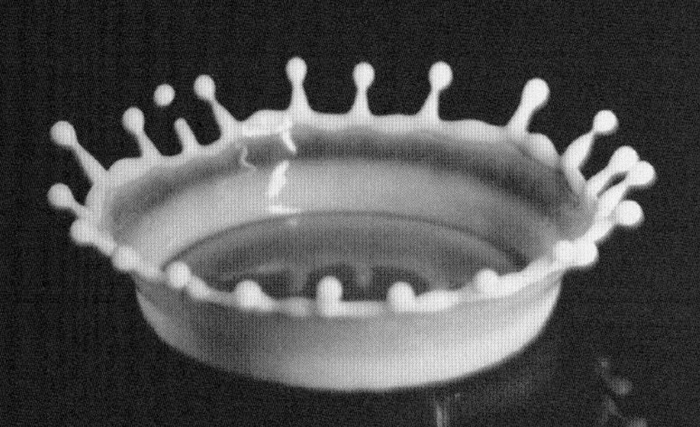

Milk-drop Coronet, 1957

Shooting the Apple, 1964

Cutting the Card Quickly, 1964

Desmore Shute Bends the Shaft, 1938

became clear, though, and in the early 1920s, when many of the firm's repairman were still in the military, Edgerton fixed electrical equipment and even strung power lines on the Nebraska prairie. Around this time, Edgerton was introduced to photography. His uncle, Frank Edgerton, ran a photo studio in Fremont, and there Harold learned how to make and print pictures.

After high school, Edgerton earned a degree in electrical engineering at the University of Nebraska, then applied to the Massachusetts Institute of Technology in 1925. Edgerton was accepted, but before classes began he was offered a place in a one-year research program to work on large electrical motors at General Electric's huge facility in Schenectady, New York. The next year he finally entered MIT, where he remained until his death in 1990.

At MIT, Edgerton studied how large surges of electricity, like those caused by lightning, affected a particular kind of electric motor. But there was a problem: The motor turned far too fast for him to see what was happening. A few years earlier, he had noticed that a thyratron—a mercury gas–filled tube that emitted regular pulses of light—had made the blades of a moving fan seem to rotate very slowly. Edgerton set up a mercury tube next to the electric motor and synchronized its pulses with the speed of the motor, so that each flash of light occurred when the motor was in the same position. It worked perfectly; when Edgerton turned the mercury tube on, the moving parts of the motor seemed to be standing still. He was able to capture this on film, using a movie camera.

What Edgerton had used was a stroboscope, a device that fools the human eye into thinking that a moving object is still. It works, Edgerton explained, because "eyes aren't designed for speed." When we see a split-second image, our brain retains it for a moment. This is how movies work: Motion picture film contains a sequence of still images, each of which pauses for a split second before the next moves into view, at a speed of 24 frames per second. A shutter blocks the frames from view as they move. The brain retains each image until the next appears, giving the impression of continuous movement. Edgerton's stroboscope allowed him to see only the "frames" created by its flashes of light. When properly synchronized with the motor, it hid its motion from view.

Edgerton knew his stroboscope could be used to examine machinery in any number of industries, not just his electric motor. He soon formed a partnership with two colleagues from MIT, Kenneth Germeshausen and Herbert Grier. The three began using stroboscopes to let businesses closely examine their

Wes Fesler Kicking a Football, 1934

Edgerton chats with Jacques Cousteau, with whom he collaborated on undersea photography expeditions.

"The trick to education," he said, "is to not let them know they're learning something until it's too late."

manufacturing processes. After a piece of equipment was filmed, the movie could be watched in extreme slow motion, and faults invisible to the naked eye suddenly became prominent.

One day, an MIT professor dropped by Edgerton's lab to see his stroboscopes and wondered what would happen if they were aimed at something besides a motor. "I suddenly realized, hey, there are a lot of things in the world that move," Edgerton remembered. "I looked around and there was a faucet right next to where I worked. So I just moved the strobe and took a picture of water coming out of a faucet." The results were amazing. Under the stroboscope, a seemingly smooth stream of water turned out to be made up of many individual drops strung together. Edgerton began aiming his camera and stroboscope at all kinds of action—the splash of a drop of milk, the wings of a hummingbird, a light bulb being smashed with a hammer.

By the late 1930s, Edgerton's pictures were becoming known to the public. (There was even a short film made about his work, *Quicker 'n a Wink*, which won an Oscar in 1940.) At the same time, photojournalists were adopting his electronic flash, using it to capture sports action and other moving subjects with extraordinary clarity.

When the United States entered World War II, Edgerton was asked to work on the problem of nighttime aerial surveillance. The army was using 50-pound bombs full of explosive powder to illuminate the night sky and take pictures of enemy territory. The bombs were dangerous and unreliable, and their use was limited. Edgerton and his partners built an electronic flash that put out far more light than anything they had made before; it could light up a square mile of land from 2,000 feet up. The flash was finally used in June 1944, to photograph the German army the night before D day. The photos confirmed that the Germans weren't ready for the massive Allied attack; after that success, the flashes were used in Europe until the end of the war.

Edgerton's next big challenge came in the late 1940s, when the company he formed with Germeshausen and Grier was asked to photograph a nuclear explosion. There was no camera capable of capturing such a detonation, which lasted only a few ten-thousandths of a second. His solution was the Rapatronic camera, whose shutter was actually a rapidly changing magnetic field that, in concert with a pair of polarizing filters, could let light in or keep it out. A magnetic field can change much faster than a mechanical shutter can move, allowing Edgerton's camera to capture exposures of a millionth of a second; its pictures revealed the glowing, bulbous form at the beginning of an atomic explosion.

But military work was only a small part of Edgerton's activities. He continued to photograph split-second action, and in the 1960s he made two of his most famous images: a bullet at the moment it sliced through a playing card, and an apple exploding on both sides as a bullet passed through it. Edgerton also applied his photographic skills to the completely different realm of the undersea world. In 1953, the famous underwater explorer Jacques Costeau enlisted his help in photographing the Deep Scattering Layer, a previously unexamined layer of living marine organisms. Later, Edgerton and Costeau used time-lapse photography to watch the graceful and surprising movements of some of the seafloor's slowest-moving creatures, starfish.

Edgerton was known almost as much for his teaching style as for his photographs. He was a firm believer in experiment, and he urged his students to learn by doing, no matter what the outcome. "Most of the things you do in life are a failure," Edgerton told them. "Then you find what the failures are and you don't do it that way the next time." His lectures were always packed, as students crowded in to see his stroboscopes stop time. This was all part of his plan. "The trick to education," he said, "is to not let them know they're learning something until it's too late."

Self-portrait with Balloon and Bullet, 1959

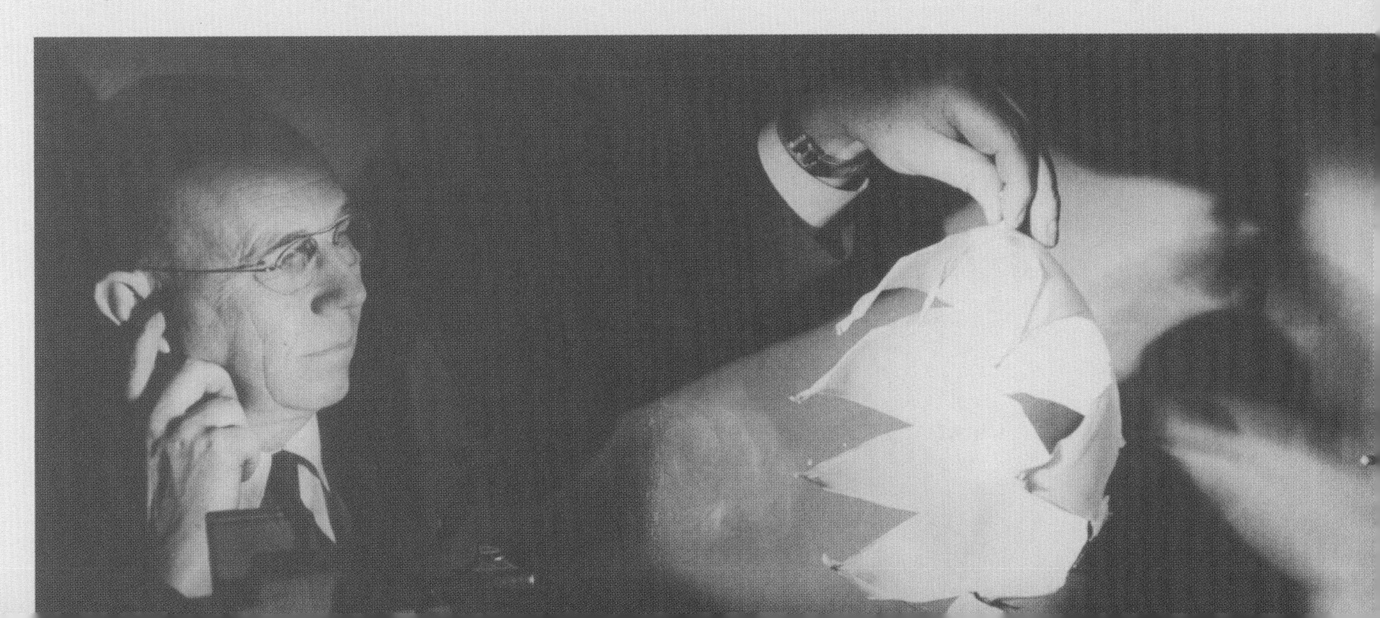

PHILO T. FARNSWORTH

Television was conceived not in a laboratory or on a university campus, but in a hay field in rural Utah. It was there, as he harvested hay row by row, that 14-year-old Philo T. Farnsworth thought of scanning and displaying a moving image line by line. Although there had been earlier attempts at broadcasting moving pictures, it is Farnsworth's system that underlies the television found in almost every home.

Since Farnsworth was not the first to think of sending moving images through a wire or by radio waves, and because years passed before he could build the system he envisioned during that harvest, it is hard to say exactly who invented television. Scientists in Europe and the United States pursued the elusive means to transmit live, moving images for almost 50 years. A German scientist discovered how to use a spinning disk to scan a moving image in 1884. An English inventor suggested using a cathode ray tube to display an electronic image in 1907. In the 1920s, researchers working independently in America and England used spinning disks to capture and display moving images. And in some ways television is simply the logical conclusion to two of the defining inventions of the nineteenth century: photography, which recorded images, and the telegraph, which transmitted information.

Still, the hundreds of millions of televisions in use today do have one inventor in common—Philo T. Farnsworth, who was born in 1906 to a Mormon family near Beaver City, Utah. In 1919, the Farnsworths moved to a ranch near Rigby, Utah. It was a decidedly rural place—every day Philo rode the four miles to the local schoolhouse on horseback. But the twentieth century had already come to Rigby, in the form of electricity. The new technology fascinated Farnsworth. He soon figured out how the ranch's power system worked and began building small electric motors in an attic workshop. And Farnsworth discovered he had a goal: to become an inventor, like Thomas Edison and Alexander Graham Bell.

In 1921, Farnsworth entered high school, but he found he wasn't challenged by his science class. He was moved to the senior chemistry class, and he began studying after school with the chemistry teacher, Justin Tolman. He learned physics and fixed radios in his spare time and read all the books and magazines about science and mechanics he could find. In 1922, one of those magazines ran a story on "Pictures That

Previous attempts at television involved spinning mechanical disks. Philo Farnsworth's breakthrough was in devising an electronic method for capturing an image, which could then be transmitted and re-created on a cathode ray tube screen.

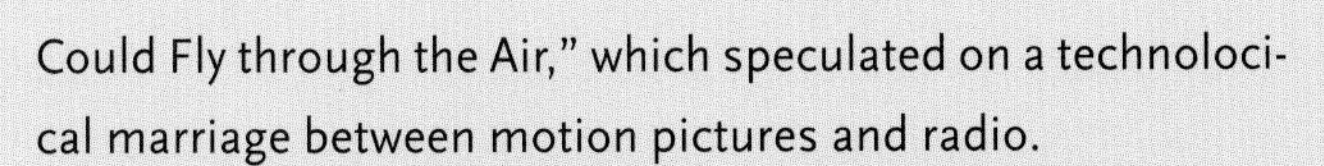

Could Fly through the Air," which speculated on a technolocical marriage between motion pictures and radio.

The story described some of the early research into television, which focused on a mechanical system invented in 1884 by Paul Nipkow, an inventor in Berlin. A spiral pattern of small holes was punched into a metal disk. When a motor spun the disk, an image in front of it would be broken up into a sequence of small parts—something like pixels on a computer screen—as light passed through the holes. The small bits of light struck a light-sensitive surface that created electric signals based on the strength of the light. The signals could then be sent through a wire to a neon bulb behind an identical spinning disk. The process then worked in reverse: the light passed through the disk's holes and formed an image on a screen on the other side. In the 1920s, two scientists, Charles Jenkins in the United States and John Logie Baird in England, worked on refining Nipkow's method. Both managed to create and transmit moving images along with sound, although the pictures were crude and

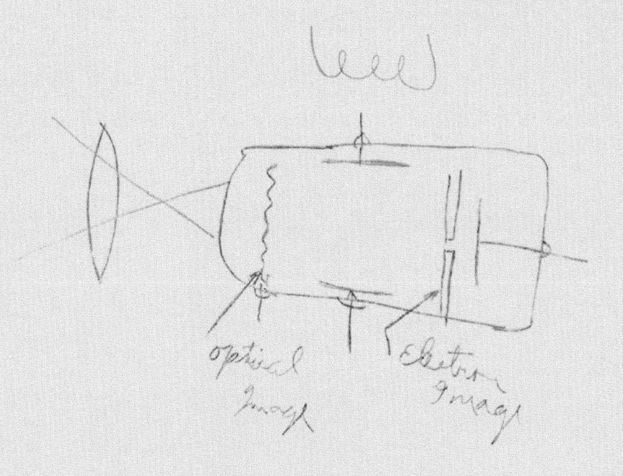

In 1934, Farnsworth and his colleagues, including secretary Mabel Bernstein, traveled to Philadelphia to demonstrate television. For three weeks, live performances were broadcast to an auditorium. Thousands of people came to see the new technology.

Farnsworth transmitted his first televised image in 1927, two years before this photo was taken. A long-running patent dispute with RCA prevented him from ever capitalizing on his technical triumphs.

fuzzy. Baird even broadcast television (or "televisor," as he called it) programs through the BBC beginning in 1929.

As Farnsworth learned more about the spinning-disk system, he was sure it would never be practical—the disks could never spin fast enough to produce a clear and consistent image. He thought of a better way to do it while he was harvesting hay on the family ranch. As Farnsworth moved his harvesting machine down straight row after straight row of hay, he envisioned a television system that worked in similarly straight lines. Electrons, he thought, could be used to scan an image line by line, then reassembled back into an image after they were transmitted.

In 1922, Farnsworth sketched out his plans for television on a blackboard for Justin Tolman. Even as he refined the scheme over the next several years, the basic idea remained. A camera lens focused an image on a light-sensitive surface; then, a beam of electrons guided by a magnetic field moved horizontally over the image, scanning it. The beam was moved down a small amount and the process was repeated. The number of electrons released by the light-sensitive surface depended on the strength of the image on it; these electrons were directed into the "image dissector," a special kind of vacuum tube Farnsworth had designed. The electrons were then transmitted to the receiver, where a similar system guided the beams of electrons, moving horizontally and then vertically line by line, to the inside surface of special kind of glass vacuum tube called a cathode ray tube. The surface was coated with a material that glowed when the electrons hit it, to create an image. Thus Farnsworth created electronic television.

In 1926, Farnsworth told George Everson, a professional fund-raiser he was working for, about his invention. Impressed, Everson staked him $6,000 to build his television. Farnsworth and his friend Cliff Gardner, along with their wives, moved to Los Angeles to get to work. It wasn't easy to build a television, they soon found. Everything had to be made from scratch; once, the Los Angeles Police Department showed up to destroy the alcohol still they thought Farnsworth was building.

The operation moved up to San Francisco the next year, and on September 7, 1927, it was ready for a test. Farnsworth, his wife, Pem, and George Everson watched the display tube in one room while Gardner placed a glass slide with a straight line painted on it in front of the camera in the other. It worked—the line was clearly visible, and when Gardner turned it 90 degrees, the first television audience saw it move. In his notebook, Farnsworth wrote, "The received line picture was evident." In a telegram to a business partner, Everson was more enthusiastic: "The damned thing works!"

Farnsworth was not the only person trying to make an electronic television, however. The man most interested in TV was probably David Sarnoff, the president of the Radio Corporation of America (RCA). Through the aggressive purchase and enforcement of patents, RCA had a virtual monopoly on the radio industry in America. With many of those patents due to expire soon and the obvious appeal of images to go along with sound, Sarnoff was eager to have a television technology of his own. In 1930, he hired Vladimir Zworykin, who had been working on the spinning-disk kind of television throughout the 1920s. Before Zworykin came to RCA's lab in New Jersey, Sarnoff sent him to look at Farnsworth's work in San Francisco. It was far ahead of what Zworykin had achieved; he is reported to have said, "this is a beautiful instrument. I wish I'd invented it." Aspects of it were to show up in his work for RCA.

Not long after Zworykin's visit, Sarnoff himself came out to Farnsworth's San Francisco lab, although only when the inventor was out of town. He was impressed, too, and offered $100,000 for the business. After Everson turned down the lowball figure, Sarnoff said, "Then there's nothing here we'll need."

Even after the depression brought America's economy to a halt, Farnsworth's and RCA's television work continued. But RCA took the competition to the courtroom as well as the lab. The corporation sued Farnsworth for patent infringement, claiming that Zworykin had invented Farnsworth's system in 1923. The hearings dragged on for months. Farnsworth's old teacher Julian Tolman was even called to testify about the drawings on the chalkboard back in 1922. In the end, Farnsworth's priority of invention was confirmed, but RCA was not easily defeated. Their appeals dragged the matter out for years, while their monopoly position let them discourage radio manufacturers from dealing with Farnsworth.

Still, there were other victories for Farnsworth. In 1934, the Franklin Institute in Philadelphia invited him to demonstrate television; for three weeks, performances by politicians, vaudevillians, athletes, and others were broadcast to an auditorium. Thousands of people showed up to watch. And in 1936, using technology licensed from Farnsworth, the German government broadcast the Olympic Games. Although it was to a very small number of receivers—and in the service of Nazi propaganda—this was the first live broadcast of a major event.

The "official" debut of television in the United States came at the 1939 World's Fair in New York, when David Sarnoff announced RCA's sets and its first broadcasts. However, RCA still didn't have any patents on its television equipment; later that year, it entered a $1 million licensing agreement with Farnsworth, the first time RCA had ever had to license another company's technology.

The spread of television was curtailed by World War II. The five-year disruption also ruined Farnsworth's chances of really competing with RCA. In 1947, just as the economy was getting back to normal, his key patents expired. That year, 6,000 TV sets were sold; by 1952, there were 22 million sets in American homes.

Farnsworth kept developing related technologies, though. He helped work on radar as well as video scanners and amplifiers. He also devoted much time and money to chasing another Holy Grail: generating power through nuclear fusion. Although he invested much time and money, nothing came of this obsessive pursuit. Farnsworth died in 1971.

STEPHANIE KWOLEK

An odd, cloudy batch of polymers that DuPont chemist Stephanie Kwolek mixed up in 1964 might have seemed like a mistake to another researcher. Usually, a polymer solution was clear and viscous. This one was thin and cloudy, almost as if it were contaminated. But Kwolek was more intrigued than disappointed, and she continued to work with the chemicals. Her diligence paid off—that milky batch of chemicals led to the development of Kevlar, a super-strong, super-stiff fiber that has saved thousands of lives.

Kevlar is now virtually synonymous with high-tech materials. Heat-resistant, five times as strong as steel, and lighter than fiberglass, it is used in hundreds of products, from protective gloves, helmets, and boots to tires, brake pads, and cables. It's used on spacecraft and to build bridges. Its most famous role, though, is in bulletproof vests, which contain several layers of a fabric woven from Kevlar fiber. When a bullet hits a vest, its thousands of pounds of force are spread out among individual Kevlar fibers. Because they are so strong and stiff, the fibers absorb a great amount of energy very quickly. According to the International Association of Chiefs of Police DuPont Kevlar Survivors Club, based in Alexandria, Virginia, at least 2,000 lives have been saved by the bullet-stopping fiber.

The introduction of Kevlar ushered in a new era of products dependent on synthetic materials. Consumers now take it for granted that materials like carbon fiber, Teflon, high-performance epoxies, and new metal alloys, as well as Kevlar, will make things stronger, lighter, and more durable. In many products, such as sports equipment or protective clothing, it's rare to find something made without new materials.

With such a big role in the invention of one of the world's best-known synthetic fibers, it's ironic that Kwolek didn't intend to become a chemist. She wanted to be a doctor.

Stephanie Kwolek was born in New Kensington, Pennsylvania, in 1923. Although her father died when she was only ten years old, he had an early influence on her interest in science. "I remember trudging through the woods near my house with him and looking for snakes and other animals," she says.

"I seem to see things that other people did not see.

If things don't work out I don't just throw them out, I struggle over them, to try and see if there's something there."

In 1964, Stephanie Kwolek investigated a curious polymer solution in her Dupont lab; that polymer eventually became Kevlar, a fiber five times as strong as steel and lighter than fiberglass. Here, Kwolek is shown holding a model of a Kevlar molecule.

The most famous application of Kevlar is in bulletproof vests, which are made of layers of fabric woven from strands of the super-strong fiber. The thousands of Kevlar fibers in a vest absorb and distribute the force of a bullet.

BULLETPROOF SPORTS EQUIPMENT

The qualities of Kevlar that make it a great material for protective gear—stiffness, strength, durability—are also perfect for many kinds of sporting equipment. While DuPont and Stephanie Kwolek may not have envisioned Kevlar-reinforced cricket bats or archery gauntlets when they introduced the material in 1971, it has proved versatile enough to work as well for a kayak as it does in a bulletproof vest. What follows is a selection of sporting equipment enhanced with Kevlar. And although many sports have already adopted the material, each year equipment designers find new uses for the super-strong fiber.

Archery

Eagle Classic Archery in England makes a 100 percent Kevlar gauntlet, to protect against the serious cuts or splinters that an errant or damaged arrow can produce.

Bicycling

Bicycle tires reinforced with Kevlar are resistant to punctures and last many miles more than traditional tires. Kevlar beads—the strips that bind tire to rim—replace steel ones, reducing the weight of the tires. Also available are Kevlar strips to mount inside regular tires, to guard against “snake bites” and other punctures.

"We also studied the various wild plants and leaves and seeds." At the same time, she adds, "I was very much interested in designing fashions. That interest came from my mother. I spent many hours creating clothes for my dolls." It was a happy coincidence that the group she later joined at DuPont was devoted to fibers and textiles.

Kwolek's mother supported the family through the depression, and when Kwolek graduated from Margaret Morrison Carnegie College (now Carnegie Mellon University) in 1946, she realized she couldn't afford to study medicine. Kwolek had majored in chemistry, and she decided to get a job in that field to save enough money to eventually go to medical school. She knew that DuPont was the leading chemical company; fortunately, Dr. Hale Charch, the inventor of moisture-proof cellophane and later a mentor to Kwolek, was interviewing prospective chemists. After some forceful insistence on Kwolek's part, Charch offered her a job on the spot, and she soon joined DuPont's Buffalo, New York, research facility.

Technical fields like chemistry had just opened up to women. During World War II, women had entered the work force in unprecedented numbers. When the war ended, many men weren't available, since they had spent the previous five years fighting, not studying. Kwolek was also lucky to sign up with Dr. Charch; under his leadership, she notes, DuPont hired more women than many other companies. Still, women were a minority in the field, and in the company. Her first promotion at DuPont came after 15 years—"far too long," she later told a reporter.

When Kwolek began working at DuPont in 1946, the first synthetic fiber, nylon, was only about eight years old. With a team of chemists called the Pioneering Research Laboratory, she began exploring new polymer fibers. Before Kevlar, she had contributed to the creation of Orlon, Lycra, and Spandex, as well as DuPont's first high-performance fiber, Nomex, which is used in firefighters' protective clothing. Kwolek also explored new ways to make these fibers, which until that time required very high temperatures. These new methods allowed her team to make and test hundreds of thousands of new polymers.

In 1964, her group decided to search for a new high-performance fiber. They had a specific use in mind for it: There were predictions of a coming gasoline shortage, and they thought that a strong, lightweight fiber could be used to reinforce car tires; cars using lighter, stiffer tires would consume less gasoline. "A number of people had been asked to take up this project and no one seemed to be particularly interested," Kwolek explains. "So I was asked if I would do it." She started out working with two "aromatic" polymers—molecules with benzene rings in them. Kwolek knew from working on

Canoeing and Kayaking

Kevlar has become one of the standard materials used in the manufacture of canoes and kayaks. It is especially useful in river and sea kayaks, which often collide with rocks. Kevlar's stiffness makes the watercraft strong, but it has more give than carbon fiber, allowing for an easier journey.

Hockey

The Kevlar in a hockey stick and blade reduces vibrations when hitting the puck, lowers weight, and increases strength.

Hunting

The Wise Hunter company recently introduced a line of camouflage hunting pants and chaps that incorporate Kevlar, to protect against venomous snake bites.

Tennis

The graphite racquet, introduced around 1980, was one of the biggest advances in the sport in the twentieth century. The addition of Kevlar to the graphite—in about a 20/80 ratio—has since increased racquets' stiffness and durability and reduced vibration.

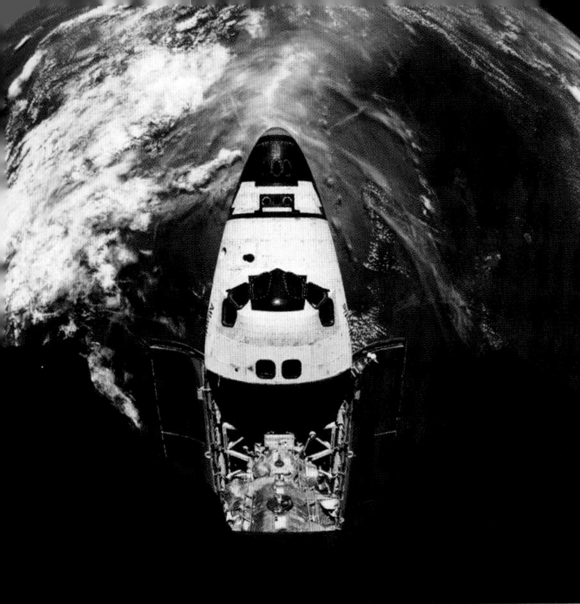

Kevlar has found hundreds of uses since its introduction in 1971, including use in space shuttle components and space suits; for suspension bridge cables; and for strengthening sports equipment such as skis.

Nomex that this kind of polymer was resistant to high temperatures and she thought these molecules might provide greater stiffness.

"In the course of that work I made a discovery," she says modestly. Under specific conditions, these polymers would form liquid crystals in a solution, which no polymer had ever done before. And instead of the usual "bent" polymer molecules—she likens them to spaghetti—these were straight, like match sticks. When the cloudy solution was "spun"—forced through the tiny holes of a device called a spinneret—the straight fibers lined up parallel to each other. This made the new fiber very stiff and very strong.

At first, though, it was not easy for Kwolek to get the polymer solution into the spinneret. "That solution was very different from the standard polymer solution," she recalls. "It had a lot of strange features about it. I think someone who wasn't thinking very much or just wasn't aware or took less interest in it, would have thrown it out." Kwolek filtered it to see if the cloudy solution was contaminated. It wasn't. Still, she continues, "when I submitted it for spinning, the guy refused to spin it. He said it would plug up the holes of his spinneret, because he assumed that [it had] solid particles. So it was a while before he consented to spinning it. I think either I wore him down or else he felt sorry for me. It spun beautifully."

The results of this first spinning were, for Kwolek, unbelievable. When a chemist spins a fiber, she sends it to a lab to test its strength, stiffness, and other properties. This new fiber came back from the lab with a stiffness at least nine times greater than anything she'd made before. "I was very hesitant about telling anyone," she says. "I didn't want to be embarrassed if someone had made a mistake. So I sent the fiber down several times. The numbers always came back in the same vicinity." Only then did she announce her results, and DuPont realized it had a fiber with great potential. After much more work and refinement by the group, Kevlar was introduced in 1971. The fiber has since found more than 200 applications, and brings DuPont hundreds of millions of dollars in sales each year.

As for Kwolek's dream of becoming a doctor, after a couple of years at DuPont she realized that the temporary job she'd taken to save some money was the right one for her. "I love doing chemistry," she says. "And I love making discoveries." Her careful, dedicated approach to research helped her succeed throughout her 40-year career at DuPont. "I'm very conscientious," Kwolek explains. "And I discovered over the years that I seem to see things that other people did not see. If things don't work out I don't just throw them out, I struggle over them, to try and see if there's something there."

It was just that will to carefully observe, struggle, and stay the course that led to her famous fiber. Today, police officers and others whose lives depend on Kevlar often come up to her to tell of their experiences. One Virginia police officer even had Kwolek autograph his bulletproof vest, which had stopped a 9 mm bullet and saved his life. "I feel very humble," she says. "I feel very lucky. So many people work all their lives and they don't make a discovery that's of benefit to other people."

JEROME LEMELSON

Few people have invented as many things, or invented for as long, as Jerome Lemelson. Over a period of more than 40 years he received, on average, one patent every month. Only a handful of people hold more U.S. patents than he, and perhaps no one in as many different fields. Many of the products we encounter every day are either based on his ideas or made with machines he helped create, from camcorders and the Walkman to industrial robots and the bar code scanner. Lemelson was also a true believer in the power of the individual inventor, devoting much of his time not only to defending his own patents but also to encouraging other people's work.

Lemelson was curious and inventive even as a child. Growing up in Staten Island, New York, where he was born in 1923, he developed a fascination with model airplanes. By the time he was a teenager, he was designing his own planes, which he flew in competitions in the still open fields of the borough. He was also learning to identify problems inventions could solve: For his father, who was a doctor, he built a tongue depressor with a built-in light, to illuminate the backs of patients' mouths.

His fascination with planes led Lemelson to enlist in the Air Force during World War II. Bad eyesight prevented him from becoming a pilot; instead, he became an engineer. After the war, he studied aeronautical and industrial engineering at New York University, where he did research for the Air Force. Between 1947 and 1951, he was part of Project Squid, which developed rocket and jet engines for improved versions of the missiles Germany used during the war.

Lemelson continued working as an aeronautical engineer after college, but he was also filling notebooks with ideas for new inventions. He applied for his first patent in 1949, for a typewriter eraser that attached to the machine with a magnet. The patent was denied—someone had already invented it. Six years later, his first patent was awarded, for a toy cap, a variation on the propeller beanie. By the end of the 1950s he had given up his engineering job to invent full time.

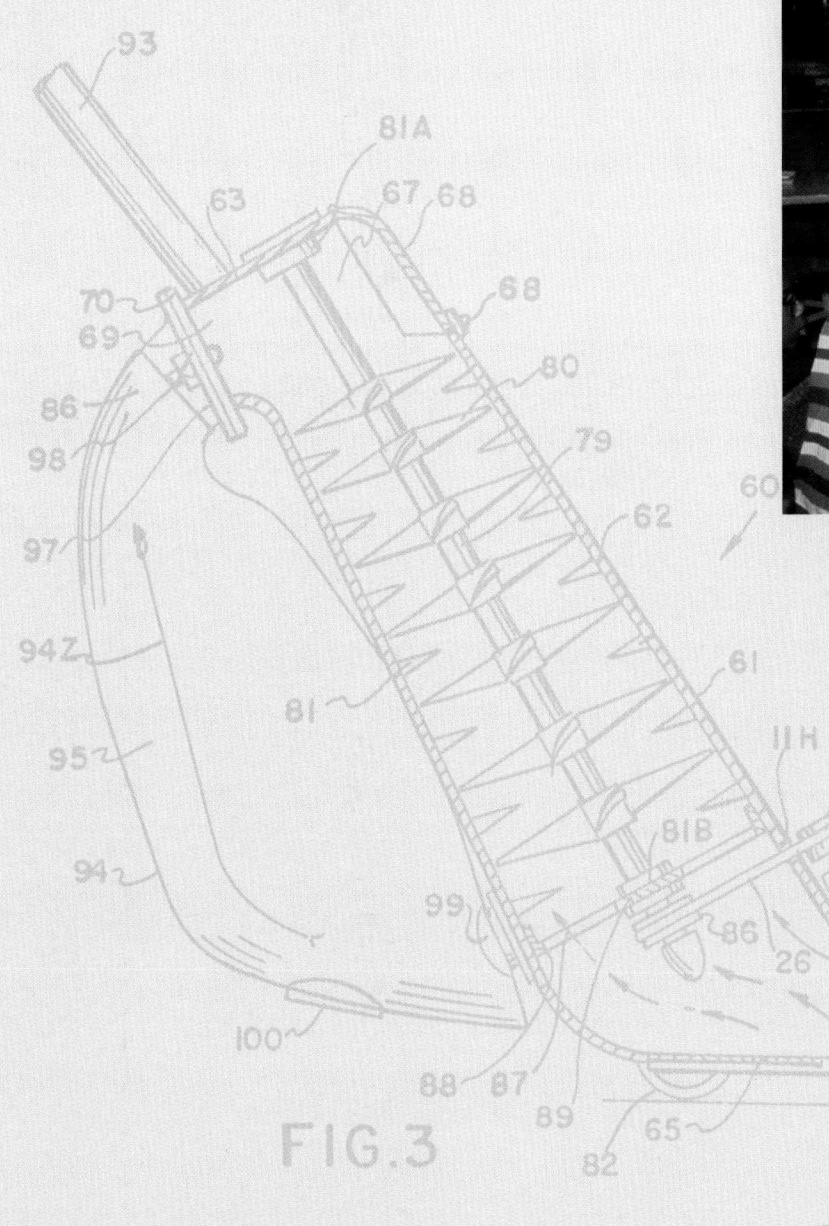

In 1995, Jerome Lemelson dedicated the Jerome and Dorothy Lemelson Center for the Study of Invention and Innovation at the Smithsonian Institution. Its mission, in part, is to encourage invention in young people and to foster an appreciation for the central role invention plays in the United States.

U.S. Patent Apr. 4, 1989 Sheet 1 of 2 4,819,101

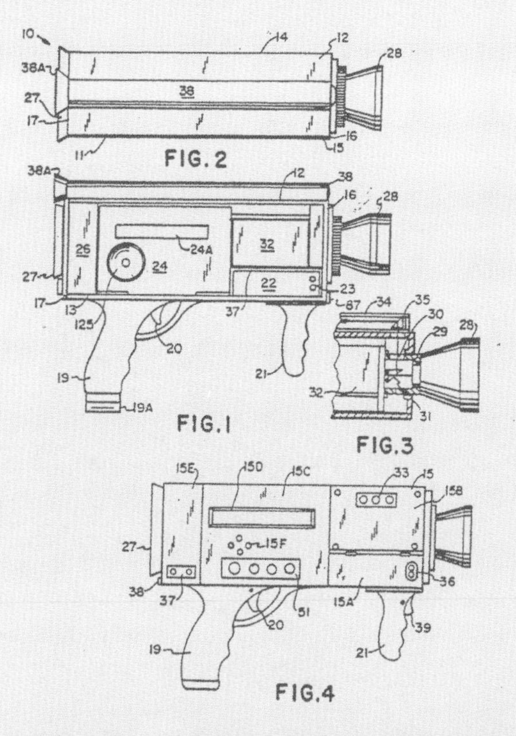

He often worked 18 hours a day, with ideas coming one after another. Regularly, Lemelson would wake up in the middle of the night to jot down an invention that had come to him during sleep. During his early years, Lemelson couldn't afford a draftsman, so he did his own patent drawings. The earnings of his wife, Dorothy, an interior decorator, helped support the family.

Many of Lemelson's early patents were for toys, since that industry depended on a constant stream of new products and was accustomed to licensing outsiders' ideas. But he was also inventing much more complex devices. The first of these was an industrial robot that could perform many tasks. Inspiration for the robot came during a 1951 demonstration of an automatic lathe that was programmed with punch cards. "I thought, 'Gee, here's a principle I can apply to many different fields,'" Lemelson recalled. In a 150-page patent application he filed on Christmas Eve 1954, he described a computer-controlled robot that

THE LEMELSON FOUNDATION

Throughout his career, Jerome Lemelson devoted time to the cause of invention and innovation in the United States. As the earnings from his many inventions accumulated, Lemelson began thinking about how to use that money to encourage American inventors and to teach schoolchildren about the excitement of invention and engineering. "We must convince our nation's young people that the field of invention can be far more rewarding—financially and in other respects—than most of them think," he declared.

In 1993, he and his wife, Dorothy, created the Lemelson Foundation to develop and fund programs devoted to invention at several American institutions. Major programs supported by the Foundation include:

Massachusetts Institute of Technology, Cambridge, Massachusetts

To celebrate excellence in innovation, the Massachusetts Institute of Technology administers the Lemelson-MIT Program, which presents inspiring inventor-innovator role models through annual awards including the world's largest for invention—the $500,000 Lemelson-MIT Prize—and educational outreach activities. The foundation also supports the teaching and training of tomorrow's innovators through the Jerome & Dorothy Lemelson Professorship and several "E-Team" courses on product development and entrepreneurship.

The Smithsonian's National Museum of American History, Washington, D.C.

Established in 1995, the Lemelson Center for the Study of Invention and Innovation is dedicated to exploring invention in history and encouraging inventive creativity in young people. The center offers lectures, conferences, publications, exhibits, educational programs, and curricular materials in both conventional and electronic form and supports oral and video history projects, fellowships, and internships. The Lemelson Foundation also supports the Museum's Hands-on Science Center, which is part of the "Science in American Life" exhibition.

Hampshire College, Amherst, Massachusetts

The Lemelson Assistive Technology Development Center (LATDC) at Hampshire College is one of the only centers of its kind in the nation. Its mission is to provide students with an experiential education in design, invention, and entrepreneurship through the use of assistive technology and universal design. LATDC achieves

could perform a variety of industrial tasks, including assembling, welding, riveting, and painting. Although "computers took up a whole room in those days," he noted, "I believed that computers would come down in size." They did come down in size, of course, and industrial robots have become standard tools in the manufacture of many products.

Lemelson's robots had to see what they were doing, and out of that necessity came his most lucrative invention, "machine vision." By converting the image from a video camera into raw data, a computer could see with great accuracy. The system could help a robot—or any other machine—tell if all of a product's parts were there, if there were defects in it, what color it was, or any number of other things. Today, this concept is applied to all automated precision manufacturing. The production of silicon chips, for example involves quickly verifying that all of a chip's connections are valid, a process that would take a person with a microscope days to do by hand. Machine vision is also used by car manufacturers and in the aerospace industry; all of the space shuttle's heat-resistant tiles, for example, are checked with such a system.

Industrial robots and machine vision were both intended to make manufacturing easier, more flexible, and more efficient. In the early 1960s, Lemelson received patents for an automatic warehousing system with the same aim. The system he outlined is familiar to anyone who has been in a modern warehouse: Computer-controlled and -guided forklifts and cranes fetch or stack pallets of goods, delivering inventory exactly when it's needed. In 1964 Lemelson had his first big success at licensing his patents when he sold the system to an Ohio company for $100,000.

While many of Lemelson's patents were focused on the way things are made, he also created products people use every

this through a combination of courses, activities, internships, and collaborations with business and nonprofit organizations, and through teams of students who design, develop, and make available equipment for people with disabilities.

National Collegiate Inventors and Innovators Alliance, Hadley, Massachusetts

The National Collegiate Inventors and Innovators Alliance (NCIIA) is an alliance of faculty and students at over 175 colleges and universities nationwide working to advance the teaching of invention and innovation in American higher education. Its mission is to nurture a new generation of innovators by promoting curricula designed to teach creativity, invention, and entrepreneurship, and to support independent teams of student innovators. The NCIIA provides grants for curriculum development and commercialization of technology-based products, as well as conferences and workshops designed to disseminate curricular innovation.

University of Nevada, Reno, Nevada

The foundation helped establish a center at the university to augment its "E-Team" program in the electrical engineering curriculum and to promote invention, innovation, and entrepreneurship in courses.

E-Teams

A common thread running through the foundation programs is "E-Teams," students working in teams to design and develop new ideas and, in some cases, to pursue their commercialization. (The "E" is for excellence and entrepreneurship.) Recent studies show that new businesses launched by multiple cofounders have a greater chance of success than those started by individuals. These results suggest that one element of academic programs aimed at preparing students for careers in business and industry should be teams of students identifying and developing a solution to a practical problem or market need, and undertaking the steps necessary to translate that solution into a commercial product.

As a young man, Lemelson was fascinated by model airplanes, which he designed and built from scratch. In this picture from the 1940s, Lemelson (right) and his brother Howard display a homemade aircraft.

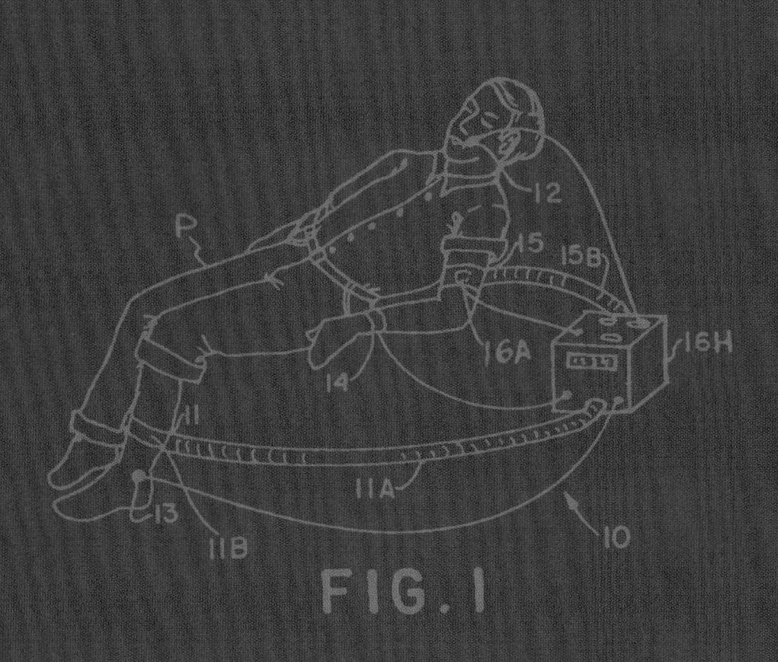

Lemelson (left) with Howard and "Pete," the dog from the "Our Gang" comedies, in Atlantic City, New Jersey, 1933.

day. Some of his patents, for example, cover technologies used in camcorders, VCRs, and fax machines. Lemelson's invention with the most far-reaching cultural impact may be his magnetic tape drive mechanism. Lemelson had first thought of using magnetic tape to store electronic images in the 1950s, when he and his wife were wading through the stacks of documentation at the United States Patent Office. More than a decade later, he created a miniature tape drive for the audiocassette format developed by Philips in the early 1960s. Sony licensed the patent in 1974 and five years later used it in the Walkman. That portable tape player changed how people listened to music and, as the first piece of electronics that millions of people carried around with them, it opened the way for the dozens of portable devices we use today, from the cell phone to personal digital assistants.

While the Sony licensing deal was straightforward, it was rarely that easy for Lemelson to control the use of his patents. He first encountered patent infringement, a problem many individual inventors face, in the 1950s. One of the toys he thought of was a mask to be printed on the back of a cereal box and cut out by kids. Kellogg's rejected the idea, so Lemelson was shocked three years later when he found masks printed on the back of Corn Flakes boxes. He sued, but at that time judges were not sympathetic to patent infringement cases, and Lemelson lost.

In the decades following his cereal box defeat, Lemelson became more vigilant in defending his patents, an effort that paid off in the late 1980s. One outgrowth of his work on machine vision had been the bar code scanner, which was part of his patent applications in the 1950s. By the time the final patents were issued in the 1980s, many manufacturers, notably automakers and electronics companies, had adopted the idea. In 1992 and 1993, Lemelson made licensing deals with most of the companies using bar codes in their factories.

The income Lemelson generated from his patents allowed him to expand his efforts at encouraging future inventors. For Lemelson, active independent inventors were the key to America's continuing industrial success. "Unless we get more of our young people interested in engineering and technology, our country will face serious problems," he said. "Engineers and designers are the front-line soldiers in any economic battle." Dorothy Lemelson adds that "he always wanted to celebrate invention. So when he was able, financially, to think about it, he sat down with [his sons], and they said, 'You've been wanting to do these things all your life, now's the time to do it. Go for it.' And he did." In the 1990s, Lemelson and his wife established the Lemelson Foundation to increase awareness of the importance of invention to the nation's economy and culture. A private philanthropy, the foundation's purpose is to stimulate the U.S. economy and secure its position in the global marketplace by creating the next generation of inventors, innovators, and entrepreneurs whose ideas will provide the basis for new companies and job creation in the twenty-first century.

Unfortunately, Jerome Lemelson was able to see only some of the results of his philanthropy before he died of cancer in 1997. Even in his last year, though, he continued inventing. In fact, the 40 patent applications he made that year alone broke even his own record for quantity, and they helped establish him as one of history's most prolific inventors. "He had an ability to take things that weren't connected at all and see the relationships between them and see what they could become," Dorothy Lemelson remembers. "I realize he saw the world we're living in [today]. In the 1950s, in his head, he saw the world that was to come."

Lemelson (right), Howard (center), and a friend experiment with an air-powered model airplane. From 1947 to 1951, Lemelson worked for the Air Force developing jet and rocket engines.

JACOB RABINOW

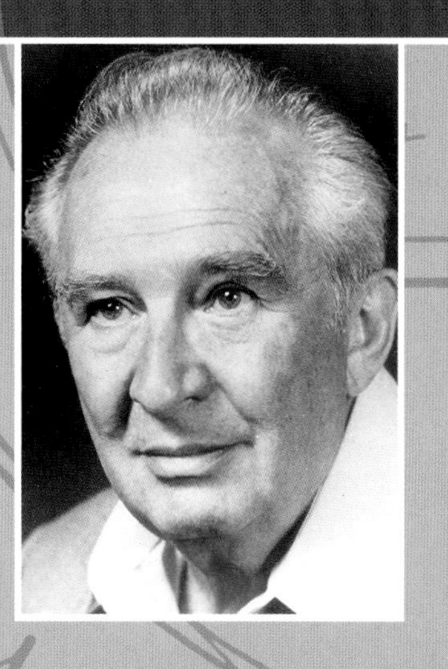

For lifelong inventor Jacob Rabinow there was as much beauty in inventing as there was in art. A successful invention, the engineer once said, "is like a good poem. . . . It does the thing it's supposed to do with a minimum number of parts."

Rabinow pursued this kind of poetry all his life, from his first invention at the age of nine until his death, at age 89, in 1999. He eventually received 230 patents, covering an extraordinarily wide range of technologies. He built the first magnetic disk storage system for computers, letter-sorting machines for the Post Office, a self-adjusting clock used for decades in most American cars, the first practical optical character-recognition system, and missile guidance systems that are still used today. There are thousands of other ideas in his notebooks, some that weren't patented because others had thought of them first (unbeknownst to him), but most because there simply wasn't enough time.

That first invention came in Siberia, where Rabinow's family had moved from Kharkov, Russia, when the threat of civil war had become too great. The Russian Revolution indeed began, and the Russian army would sometimes come through his Siberian town. Rabinow was keenly interested in their equipment. (He was also an avid reader of Jules Verne's science fiction.) One day, Rabinow and his brother made a small machine that could throw rocks, building it out of a couple of posts and some twisted rope. It worked, but an adult gave the young inventor some bad news: He had reinvented a Roman *ballista,* a 2,000-year-old technology. ("I didn't know there were Romans," he later said.)

Rabinow's father died of typhus not long after the Communists closed down his shoe factory. The family fled to China in 1919, and from there moved to Brooklyn, New York, in 1921. Rabinow's mother worked making corsets, and he attended public school, where he continued to be fascinated by engineering and dreamed of becoming an inventor. But people told Rabinow that he would never become an engineer in America because he was Jewish. He listened to them, and he graduated from the City College of New York with a degree in "straight" science. Rabinow wasn't satisfied with this for long, though; after graduation, in the middle of the depression, he decided, "If I'll starve, I'll starve doing something I like." With some borrowed money and savings from working in a radio shop, Rabinow went back to City College and earned a masters degree in engineering.

In 1938 Rabinow got a job at the National Bureau of Standards (now the National Institute of Standards and Technology), performing the mundane task of calibrating water meters. His inventiveness was soon noted by his superiors, and as America began to prepare for World War II, Rabinow started working on something much more interesting: fuses for bombs and missiles.

An armed bomb is a dangerous thing; the trick is to activate it when it's safely away from the people firing it. The device that does this is called a fuse. Although the technology for arming conventional shells already existed, missiles were new, and the old techniques didn't work. Rabinow's first fuse consisted of four sets of weights. To activate a missile, the force of acceleration would move the weights one after the other, until a circuit was completed. If the missile was simply mishandled or dropped only one set of weights would move, and the missile would not arm.

It was much more difficult to make a fuse smart enough to arm and detonate a missile only when it was close enough to its target to damage it. (Earlier proximity fuses, as they're called, worked either on contact with a target, which limited the damage they could do, or with timers, which often made them explode too early.) Rabinow's major contribution to proximity fuses was to realize that the shock wave created by an aircraft traveling faster than the speed of sound could be used to trigger a missile. Later, he was asked to work on the problem of keeping a missile's radar antenna straight even as the missile wobbled. A modified version of the gyroscope mounting system he invented is still used in Sidewinder missiles.

Rabinow was a constant, even compulsive inventor. One of his favorite sayings was, "We all curse at first, but inventors fix." Any mechanical challenge or shortcoming at work or home was grounds for working on a new

Rabinow received some 230 patents during his life, ranging from industrial devices such as automatic mail-sorters and character recognition systems to pick-proof locks and a new method for hanging pictures.

THE QUOTABLE JACOB RABINOW

Jacob Rabinow loved talking about his inventions and about invention in general. He returned again and again to certain themes, such as the moment of invention, the art of invention, and the importance of cultivating American innovation. Here are some samples of Rabinow's wisdom.

On inventing

"People are so carried away with their own brilliance that it is hard for them to believe that the world is filled with equally brilliant people. One gains tremendous respect for the human race when one sits in the Patent Office Search Room for a day and looks through the tremendous variety of ideas on any subject one can possibly conceive."

"If you want to be different, you better be good. If you want to make a really different product, it better be very good."

"The feeling I get when I see [an elegant device] is exactly like that of a person who likes art and sees a great painting for the first time."

"Inventing is a hell of a lot of fun if you don't have to make a living at it."

On working without official permission

"Things that are done illegally are done efficiently."

On designing weapons

"You may wonder how I can talk so coldly about designing weapons, throwing bombs, and killing people. . . . I think that of all the things mankind has done and is doing now, the most obscene, the most vile, the most terrible, the most cruel by far is war. And yet, I am not a pacifist, even though I had seen war and death when I was a young boy in Siberia and know what it is like to be terribly frightened. I think it is better to kill than to be murdered in cold blood. I think it is necessary to fight despotism and sometimes it is necessary to fight in self-defense."

On invention in America

"Invention is culturally driven, and American culture today is not a strong driver. If a country wants great inventors, it has to make sure that inventors are encouraged, rewarded, liked, made heroes."

On the joys of inventing

"It would be wonderful if all inventors started by being very rich, so that selling or licensing a patent would make relatively little difference to them. Unfortunately, that does not happen often. Although I have known one or two inventors who were very rich, I have met thousands who aren't. They were not rich in money, but there are other riches. There is excitement in seeing a glimpse of the future, leaving something better than you found it, of leaving a scratch on history."

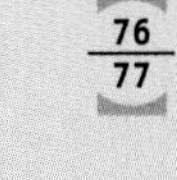

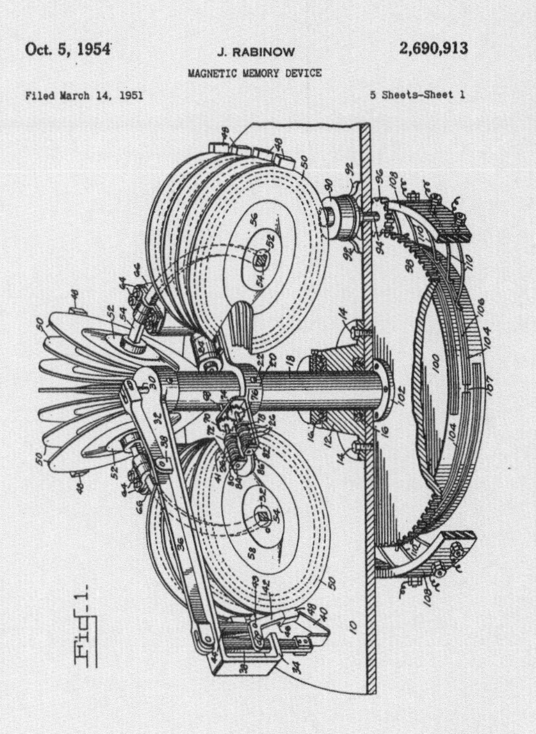

Rabinow's magnetic memory device was a precursor to the modern hard disk drive, featuring spinning, magnetically charged disks that held data.

device. When a short friend complained that her feet didn't reach the ground from a Washington, D.C., theater seat, Rabinow whipped up a portable stool whose legs folded up into a modified hardcover book. At home, he noticed how hard it was to read the label on a rotating record album, so he created a spinning prism device that would make the label seem to be standing still. He even created a new way to hang pictures that wouldn't damage his walls.

While these fixes for everyday life pleased Rabinow, he was even better at envisioning solutions to complicated problems. In 1945, his wife, Gladys, gave him a Swiss watch as a birthday present. Although it was an excellent watch, like all timepieces made then it slowed down after a while. Rabinow opened up his new watch—scratching its nice case in the process—and discovered a small lever that would speed up or slow down the watch. He adjusted the watch, but a year later it slowed down again. Although watchmaking had been one of the most advanced crafts skills for centuries, no one had figured out how to make a watch that could automatically adjust itself. Rabinow set to work, and made a very small limiting device that adjusted a watch's speed when the time was reset—a watch owner would never even know he was fixing the speed.

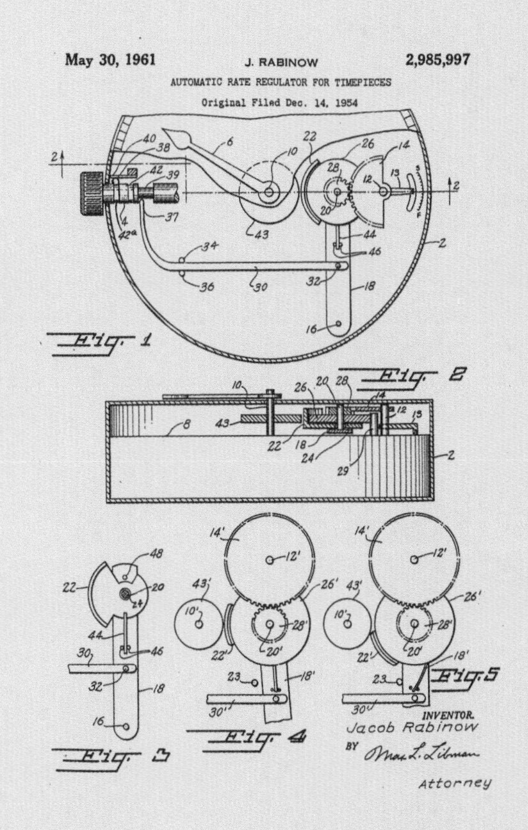

Rabinow's first moneymaking invention was the self-regulating clock, which adjusted itself as it slowed down. Watchmakers didn't want to admit their products were less than perfect, so Rabinow sold the idea to carmakers.

Jacob Rabinow always enjoyed discussing inventions, and especially his inventions. Here he illustrates the principles behind the "magnetic memory device" he invented in 1951.

One of Rabinow's inventions was a new kind of Venetian blind—no design problem was too small for him to address.

The watch industry did not accept Rabinow's advance easily. To call a new watch "self-adjusting" implied that older watches weren't as good—which they weren't. Eventually, Rabinow's device was licensed to automobile clock makers, since it was almost impossible to get to the workings of a car's clock to adjust it. This turned out to be Rabinow's most lucrative invention—he netted one cent for every clock made, about $20,000 a year.

Rabinow was particularly intrigued by problems others thought were impossible. "I had once heard the U.S. Postmaster General say that the Post Office could not be automated," he recalled in his 1990 book *Inventing for Fun and Profit*. "I was willing to bet lunches for a year that something could be done." He had already done some work for the Census Bureau designing a punch-card sorting machine and thought he could translate some of that experience to the task of sorting millions of letters a day. The system he invented was complicated, involving custom keyboards, sorting trays with built-in mechanical "memory," vacuum devices to lift the letters, and intricate "roller coasters" to move the trays. And it worked; Rabinow's system, finished in the 1960s, was only recently replaced by newer technology.

A key aspect to sorting mail was recognizing the often irregular letters in the address. After working on a system for automatically selecting microfilm documents, Rabinow realized he could make a machine that could read—or at least recognize individual letters. A character was projected onto a piece of film; behind that film was a spinning disk printed with each character of the alphabet. By means of light-sensitive photocells, each character was compared with the ones on the disk, until a match was found.

The secret to Rabinow's Optical Character Recognition (OCR) system was that it didn't try to match each letter exactly. "If a character is printed perfectly," he wrote, "any reading system works." But when a character is poorly printed, it's nearly impossible to come up with a perfect match. So Rabinow's machine used a "best match" approach, deciding that a character was an A, for example, because it was more like an A than anything else. The system wasn't 100 percent accurate, but it was close, and far better than anything else on the market. Although the first OCR system was very slow—it read a character a second—refinements eventually allowed it to read up to 14,000 characters a second. The "best match" system is still used for character recognition, and the principle has been successfully applied to other fields.

Rabinow's belief in the power and importance of invention was not limited to his own work. He seized on any opportunity to talk about innovation and the state of invention in the United States, which he was very worried about. In the 1980s, he contributed several articles to scientific magazines that not only detailed his own approach to invention ("You have to realize that invention is an art form") but also declared the need for a better attitude toward innovation and small inventors. "Let's support the small inventor," he wrote, "not because he's going to save the country, but because he will keep the big guys honest." Although he officially retired from the National Institute of Standards and Technology in 1989, he continued to advise the agency and helped it plan a museum of American technology. He also continued to invent, and he spent much of his later life solving a pet problem, one first posed to him by an NBS colleage in 1938: creating a pick-proof lock.

PERCY SPENCER

The microwave oven began with a sticky mess in the pocket of Percy Spencer's lab coat. One day in 1946, he was testing a magnetron—the tube that powers radar systems—at the Raytheon Company, where he worked. When he reached into his pocket for the chocolate bar he had been carrying around, he found it had melted. Although many people would have assumed that the chocolate had melted from body heat, Spencer guessed that the culprit was microwaves emitted by the magnetron.

To test this hypothesis, Spencer sent an assistant out for a bag of popcorn. When the kernels were held up in front of the magnetron, they exploded all over the lab. The next morning, Spencer conducted another experiment. He took an old tea kettle and cut a hole in its side. Then he put an uncooked egg in it, aimed the magnetron at the kettle's hole, and powered it up. A skeptical colleague peered into the kettle just as the egg exploded—the yolk had gotten so hot that its steam burst open the shell. By 1947, Raytheon had built the first microwave oven, and in 1949 Spencer received a patent for his new "method of preparing food."

During his youth, Spencer would have seemed an unlikely candidate for inventive greatness. Born in 1894 in the small town of Howland, Maine, Spencer's early life was full of tragedy. His father died when Spencer was just 18 months old, and his mother left home soon after that, leaving him to his aunt and uncle. When he was seven, that uncle died, too. The small family's modest circumstances prevented Spencer from completing school; when he was just 12, he began working at the local spool mill.

When he was 16, Spencer took a job that set him on a different course. The local paper mill was buying its first electrical system, and although the teenager knew nothing about electricity, he was hired to help install it. Within two years, he was a full-fledged electrician, having learned by trial and error. In 1912, inspired by the story of how a wireless radio operator had helped save many passengers on the *Titanic*, Spencer joined the navy to learn radio. After his tour of duty (and an education at the Naval Radio School), he joined a company that was making radio equipment for soldiers fighting World War I.

Barely 21 years old, Spencer threw himself into the work, logging long hours not only building radios but also trying to figure them out. "Many's the time the gang would come back in the morning and find Percy still there," a coworker recalled to *Reader's Digest* in 1958. "He had stayed up all night just to find out how things worked."

In the late 1920s, Spencer joined Raytheon, which had been founded a few years earlier by Vannevar Bush and two other scientists. (Bush would later lead the Manhattan Project, which built the first atomic bomb.) Although the company's first product, a refrigerator, was a flop, it soon became known for its new invention, the radio tube. Through the 1920s and 1930s, Spencer worked on new kinds of radio tubes, often in collaboration with physicists at the Massachusetts Institute of Technology. "Spencer became one of the best tube designers in the world," one of those physicists told *Reader's Digest*. "He could make a working tube out of a sardine can." One of his tubes eventually became part of the modern television camera.

When World War II began in Europe in 1939, Raytheon, like many companies in the United States, shifted its focus to military work. In 1941 Spencer began working to improve a top-secret project of the British: radar.

Radar, short for "radio detection and ranging," was an extremely important invention. It allowed the British to "see" enemy planes and ships from a great distance, even at night or in bad weather. The idea behind the technology is fairly simple. Microwaves, which are electromagnetic waves shorter than radio waves but longer than the waves that make up visible light, are transmitted into the air. When they hit an object, such as a plane, they are reflected back. An antenna picks up these reflected waves, and by measuring the time between transmission and reception, one can determine exactly where the object is. Repeating the process reveals what direction the object is moving in, and how fast.

The heart of a radar system is the magnetron, a tube that produces and emits microwaves. The magnetrons designed by the British were "awkward and impractical," according to Spencer. Although thousands of them were

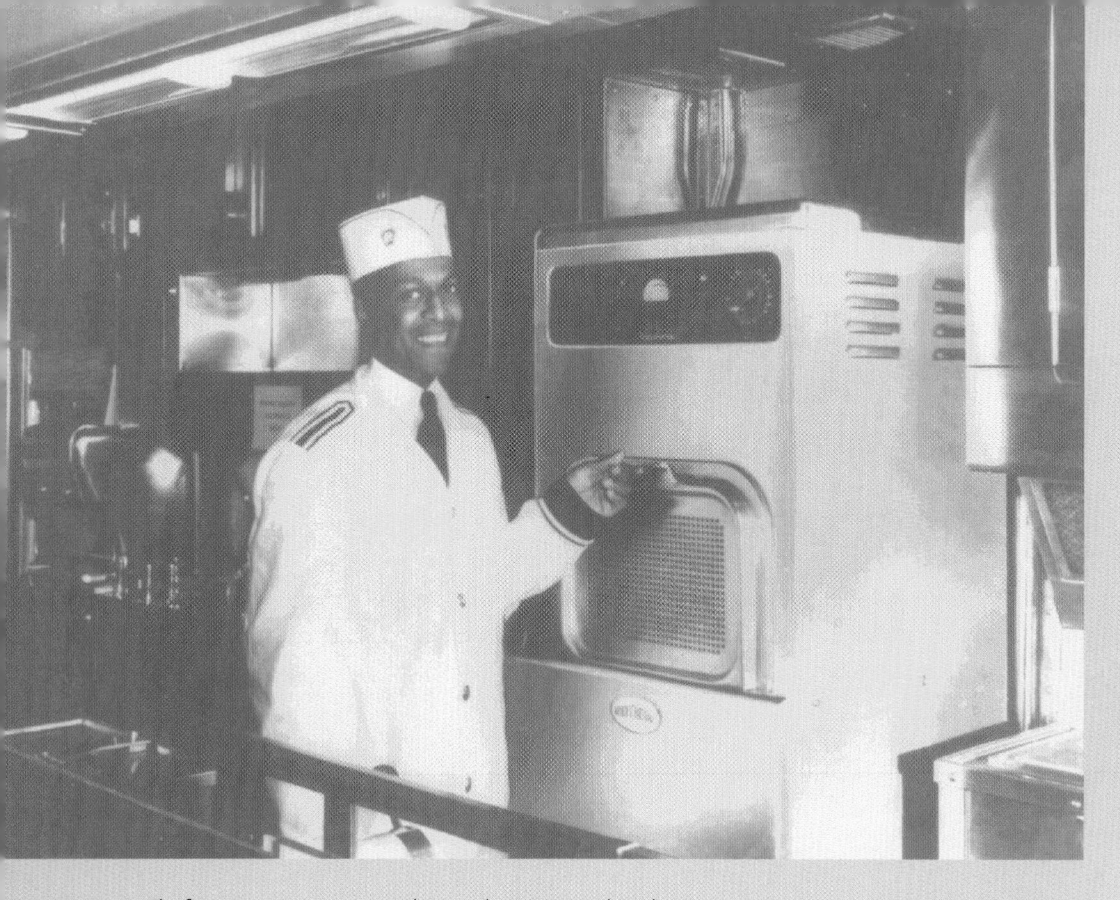

The first microwave ovens were large and expensive and used mainly in institutional kitchens: cruise ships, resorts, and trains.

The core of a microwave oven is a magnetron (sitting on Spencer's desk), a tube that creates a high-frequency electromagnetic field.

needed, it could take a machinist a week to build just one magnetron. Spencer was given the task of speeding up this process. Using the British design, his workers at Raytheon were able to make about 100 a day. Still, that wasn't nearly enough. So Spencer redesigned the magnetron, creating a model that was much easier to build. Soon, Raytheon was producing 2,600 magnetrons a day.

Spencer's redesigned magnetron also led to more sensitive radar systems; when his system was installed in U.S. bombers, it could detect an object as small as a German submarine's periscope. For his radar work, the navy awarded him the Distinguished Public Service Award, its highest honor for a civilian.

This extensive experience with magnetrons let Spencer quickly see why his chocolate bar had melted. And while other scientists had noticed that microwaves could make things heat up, only Spencer thought to use them to cook food. It was a conceptual leap, since microwaves heat food in a completely different way than conventional ovens.

Normal cooking works by thermal conduction—the heat produced by gas, electricity, or fire starts at the outside of the food and moves in, "conducted" by the food itself. Microwave cooking, on the other hand, does not begin with heat. Microwaves move freely through food (and many other substances); as they do, they excite the molecules in water and fat. The excited molecules vibrate, which produces friction between them. That friction, in turn, produces heat, which actually cooks the food. (It's a myth that microwaves cook from the inside out; they simply cook everything at once.)

Raytheon's first microwave oven, called the "Radarange," was huge and primitive—essentially a large, liquid-cooled magnetron pointed directly into a metal box. About the size of a refrigerator, weighing hundreds of pounds, and priced at $2,000 to $3,000, the few that sold went to restaurants, the military, and railroads. In the 1950s, Raytheon and the Tappan company produced a smaller model for $1,295; this also failed commercially.

Success came only after Japanese engineers had shrunk the magnetron to a size that would fit in a small box. In 1967, the Amana appliance company (which Raytheon had bought a couple of years earlier) introduced the first compact microwave oven. At $495, it was affordable for many families. And although the new methods of cooking frightened a lot of people, by 1975 microwave ovens were outselling gas ranges.

While microwaves may never replace conventional ovens, Spencer's accidental invention has made its way into some 90 percent of the kitchens in America.

TRANSPORTATION

INTRODUCTION

Although the quote is often attributed to Isaac Newton, it was in fact the thirteenth-century English cleric and noodler Roger Bacon who said, of innovative thinkers like himself: "We stand on the shoulders of giants." In different ways, this group of change-makers represents that view of invention: What comes from their endeavors is only the latest improvement in a long historical series of gradual development. This is not to denigrate in any way the essential value of any single innovation. Far from it. In one sense, the latest high-tech plasma torch or electrophoretic technique is part of a continuum that started with the first flint tool. No invention comes, like Minerva, fully formed from the brow of Zeus. Each represents the final piece of a local section of a jigsaw puzzle. Even the great Thomas Edison might have failed in his light-bulb efforts had it not been for a newly developed air pump from the lab of German chemist Hermann Sprengel. That pump made possible the high vacuum inside Edison's bulb on which the consequent longevity of his burning filament depended.

The innovative continuum reaching back into history is comprehensive in scope. One example of a sequence taken from it will illustrate the point: The cams used in medieval gearing mechanisms (working mills of all kinds) were later adapted as pegs on the surface of a cylinder spun by waterpower to trigger devices that would automate the motion of automata or control and direct the flow of fountains in Renaissance water gardens. One of the most famous was at the Villa d'Este in Tivoli, outside Rome, where cam-controlled machinery caused statues to move, music to be played, and fountains to weave fantastic patterns in spray. Failure of the pumps used to lift the well water required for the gardens led to investigation by such luminaries as Galileo Galilei and his pupil Evangelista Torricelli, whose subsequent discovery of air pressure and then of the vacuum would eventually make possible James Watt's improved steam engine, which used the combination of vacuum inside a cylinder and air pressure outside it to move the cylinder piston up and down. The steam engine in turn made possible the locomotive, in one form developed by the Swedish inventor John Ericsson, who went on to fame (if not fortune) as the designer of the *Monitor* and the first American ironclads. By the time these had developed into late-nineteenth-century all-metal ships, their navigators were beginning to have problems with the effect of an iron hull on their compasses. This problem was solved through the adoption of the gyrocompass, which depends for its accuracy on inertia rather than magnetism. And by a touch of the serendipity with which innovation is rife, the gyrocompass was developed by Elmer Sperry (who appears in this book), one of the many inventors whose technology was a direct descendant of the medieval millwheel.

This group of inventors, however, shares another, particularly American, aspect. All belong to a "can-do" American tradition first formally noted in a British Parliamentary Committee report of 1841, which warned parliamentarians of the real danger that one day the mighty British industrial machine would be overtaken by that of the United States, thanks to "the American system of manufacture," of which Henry Ford is a prime example. From the early years of the Republic, without the restrictive old European craft practices to hold them back, Americans embraced the use of labor-saving machinery as did no other Western nation. They were encouraged in this by a scarce labor market. When most male immigrants headed away from the coast to buy cheap land and become farmers, the only readily available labor pool left in the East was made up of women without craft experience. So, machines were developed that could operate in semiautomatic mode, requiring only unskilled (female) machine-minders.

This readiness to adapt to circumstance is one of America's most enduring characteristics and is what makes the American social environment more amenable to innovation than any other, because America sees change as an opportunity rather than a threat. This attitude is rooted in America's past. The frontier was no place for the suffocating grip of accepted practices. Self-help was the only way out of most local problems, and resulted in such great inventions as barbed wire.

The Victorian writer and reformer Samuel Smiles made the innovative loner a hero, less in his native Britain than in America. Smiles's biographies of engineers helped create the role model of the individual problem-solver working to improve the daily life of ordinary people. In the nineteenth century great innovative effort was expended on matters relating to transportation, as the country grew and expanded: railroads, automobiles, bridges. Much in the same vein, in the twentieth century this group of inventors' efforts ranged from essential but everyday necessities like the traffic signal, to exotic examples of derring-do such as the solar-powered airplane, to the grand scale of the interplanetary launch vehicle.

This group of inventors also belongs to the tradition, born of the American system of manufacture, of helping to provide society with a democracy of possessions. Ford took a production-line concept originating with the British Royal Navy (and at the time not taken up elsewhere in British industry) to make automobiles that would be cheap enough for everybody to own. Ole Evinrude upgraded ordinary people's leisure time with an affordable outboard motor. By such unglamorous advances is the quality of life enhanced.

—JB

OLE EVINRUDE

On a hot August afternoon in 1906, Ole Evinrude went for a lakeside picnic with some friends and his fiancée, Bess Cary. The party had rowed two and a half miles out to a small island in the Wisconsin lake when Bess said that she'd really like some ice cream. Evinrude, already devoted to his future wife and business partner, jumped back in the boat and rowed to shore to fetch some. As he struggled in the heat on the five-mile round trip, he had an idea: Someone should invent a motor for his boat. That someone, it turned out, was Evinrude himself. He already had experience building small gasoline engines, and by the next summer he had built his first outboard motor, the prototype for hundreds of thousands of motors that have changed the face of aquatic transportation.

Until that moment of ice cream–inspired invention, Ole Evinrude's life had been typical of the late-nineteenth-century immigrant experience. The first of his parents' 11 children, he was born in 1877 on a small farm in Norway, about 60 miles from Oslo. His father was a practical and frugal farmer; his mother, who tended the farmhouse and the children, was descended from a family of blacksmiths. Evinrude's earliest memories from Norway were, appropriately, of local marine activity. "What I remember first in my life is the lake nearby," he said later, "playing on its shores and watching the boats."

When he was five years old, Evinrude's family moved to America. His second memory is of that voyage from Norway: "My mother and grandmother were constantly fishing me out of the engine room, only to have me dash back again. I couldn't possibly stay away from the engines." The family settled near a lake in Wisconsin, where there was already a sizable Scandinavian community. The young Evinrude helped out on the farm his father was homesteading and also attended two local schools—one taught in English, the other in Norwegian. But he stopped going after the third grade; he was needed on the farm, and he was also far ahead of his classmates. "I could do all the problems in the book up to the eighth grade," he recalled.

As Evinrude grew up, his thoughts remained with boats and engines. One of his uncles was a sailor, and on visits to the family he taught the boy how to distinguish the different kinds of ships. With money saved up from a job sorting tobacco, Evinrude subscribed to a magazine devoted to science and mechanics. Before he was 16, Evinrude tried to build his own sailboat, carving its parts out of scrap lumber. But when his father found the pieces in their woodshed, he threw them into the fire, feeling that they were a waste of money and time. Undaunted, Evinrude started his boat over, this time hiding the parts around the farm. When his father was away for a few days, a little after Evinrude's 16th birthday, he assembled them into an 18-foot sloop.

This time, his father wasn't angry. He recognized his son's dedication and skill, and (after Evinrude began charging tourists a quarter for a sail on the lake that summer) his business sense, too. When the season was over, Evinrude took his earnings and set out for Madison, where he apprenticed at a farm machinery shop. With the job paying just $2.50 a week and room and board costing $3.50, his boating earnings turned out to be essential.

For the next several years, Evinrude worked in different kinds of factories throughout the industrial Midwest. He toiled in a steel rolling mill, made machine tools, and began working with the then new gasoline engine. In 1900, at the age of 23, Evinrude returned to Wisconsin, taking a job as a patternmaker at a Milwaukee machine shop. At the same time, he was carrying on his own experiments with internal combustion engines, even firing one up in his boarding house room. Evinrude's landlady forced him to relocate his workshop to a nearby shed, where he managed to build a horseless carriage. Local accounts described that contraption as a "fearsome thing, with a bark like a sea lion and an exhaust that fogged the whole street with smoke."

Bess Cary lived next door to Evinrude's shed workshop, and the two developed a friendship. When Evinrude and another man founded a company to manufacture small engines, Bess typed their correspondence. The engine company was successful at first, selling 50 portable engines to the government. (It was those portable engines that helped Evinrude see the possibility

Evinrude's first motor was a simple machine: a small gasoline engine was mounted on top of a driveshaft that turned a propeller. A standard clamp attached the motor to the back of any boat.

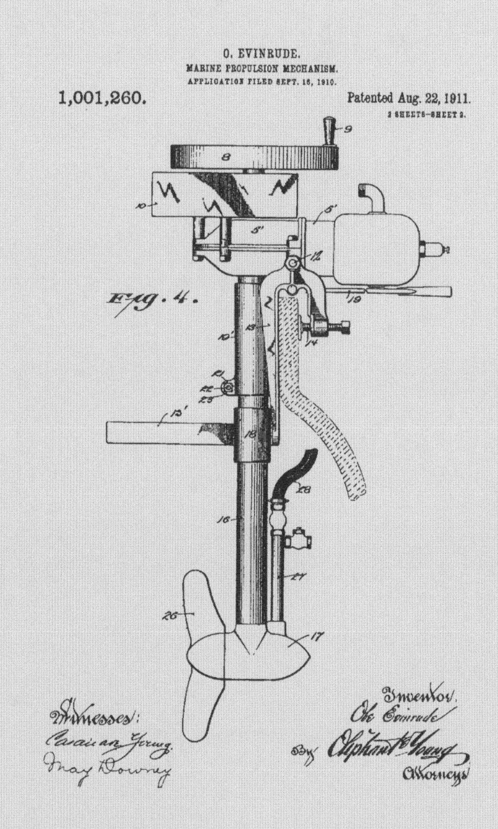

of a small motor for his row boat.) Bess and Ole did even better: In November 1906 they were married.

Evinrude's first company failed within a year or two, and in a very short time he went through two other failed motor companies. With a wife (and soon a son, Ralph) to support, Evinrude went back to the more reliable occupation of patternmaking, with Bess once again taking care of the business correspondence. He was soon running a successful pattern shop, yet in his few spare evening hours (and even when he was bed-ridden with rheumatism) he would turn to his idea for a portable boat motor.

The design Evinrude came up with was simple: A single-cylinder gasoline engine turned a long, slender driveshaft, at the bottom of which was a propeller. A clamp would hold the motor to the flat back end of a boat. In the summer of 1907, he proudly showed his first motor to his wife. She chastised him for wasting the family's money on such a thing and declared that it looked like a coffee grinder.

Bess Evinrude also saw the potential of her husband's invention, however. She encouraged him to refine the design, and when one of Evinrude's employees put that engine on a boat the next summer and put-putted around a local lake, her intuition proved right: He returned with orders for 10 of them.

Evinrude was not the first person to think of the outboard motor. In 1896, a New York company sold 25 "portable boat motors," and in 1905 a Detroit inventor, Cameron Waterman, began building his own motors, selling some 3,000 of them in 1907. But Evinrude's motor was more reliable, simpler to build, and the first to be well promoted.

Evinrude began producing his motor in larger quantities in 1909, and by 1911 had entered into a partnership to found the Evinrude Detachable Row Boat Motor Company. Perhaps more important, in 1909, he advertised the motor for the first time, with a slogan written by Bess: "Don't Row! Throw the Oars Away! Use an Evinrude Motor." The ad was perfectly suited to the era. With new technologies such as electricity and the automobile, it was finally possible for people to be free of the drudgery of manual labor. And as Evinrude had learned on that quest for ice cream, there was considerable boating labor to be freed from.

The ad campaign went national in 1911. That year, the company sold about 1,000 outboards; 4,650 were sold in 1912, and 9,412 in 1913. It seemed that Evinrude had found some permanent industrial success. But the effort to keep the business go-

Inspiration for building an outboard motor struck Ole Evinrude on a hot summer day, when he grew tired of rowing to an island where he was picnicking.

ing had taken a serious toll on Bess Evinrude's health, and in early 1914, Ole sold his share of the company to his partner. Bess, Ole, and Ralph drove around the United States, settling in Florida and then New Orleans, where her health slowly improved.

Although Evinrude had agreed not to compete with his company for five years, by 1917 he was designing a new and better outboard motor. When the five years were up, he offered his new design—a lightweight two-cylinder engine made of aluminum—to Chris Meyer, his former partner. Meyer didn't want to give up on the proven success of the old model, however, and declined to buy Evinrude's new design. It was a fatal mistake; Evinrude started another company, Elto, and in 1920 began selling the new motor. It was a quick success, while the older Evinrude motor slowly lost market share.

The Elto engine had to compete with more than its Evinrude ancestor. In the five years of Ole Evinrude's exile, the outboard market had changed. Speed was becoming more important than simply getting around, and the new Johnson Motor Company's engines could propel a boat at up to 16 miles per hour, versus the Elto's 10 mph. The three companies were highly competitive, with Johnson and Elto continually refining their motors. In 1929, Elto merged with the old Evinrude company and another competitor to form the Outboard Motors Corporation (OMC). Six years later, during the depths of the depression, OMC bought Johnson, consolidating the industry.

In 1933, Bess Evinrude passed away, ending not only a marriage but also Ole Evinrude's most successful business partnership. He died a year later, a few months before his son orchestrated the purchase of Johnson. Ralph Evinrude led the company through the depression and on to greater success. Today, the Evinrude and Johnson brands live on as part of the Boats and Outboard Engines Division of Bombardier Motor Corporation of America, almost a century after its founder first grew tired of rowing.

When Evinrude's wife Bess fell ill, Evinrude left the outboard motor business. Idleness did not suit him, though, and he designed a lighter and more powerful motor. When Evinrude's original company rejected the motor, he founded a new firm, Elto.

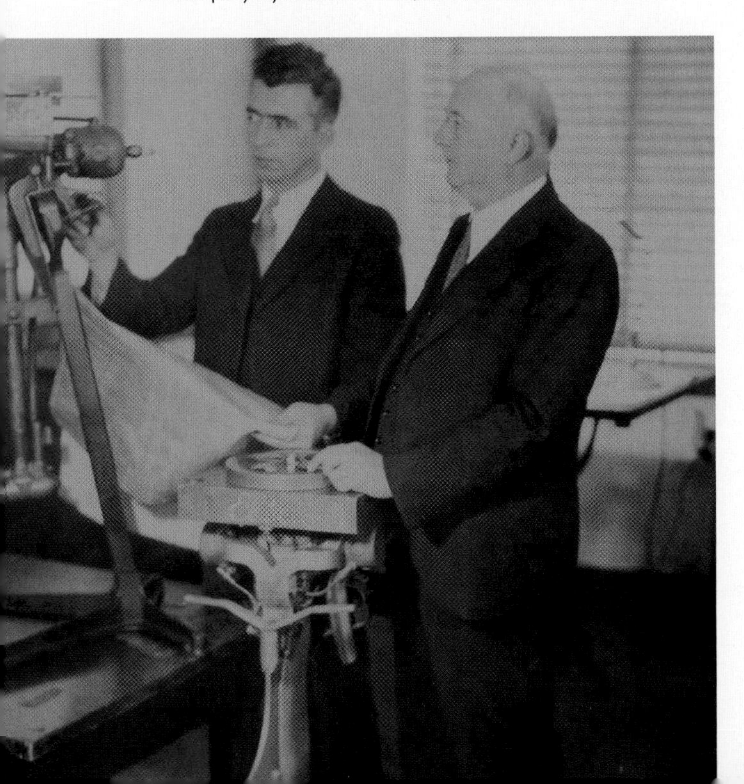

HENRY FORD

Henry Ford wasn't the first person to build an automobile. He was not even the first person to build an automobile in Detroit. Ford's genius wasn't in being there first, but in doing it best. The car that he introduced in 1908, the Model T, was better engineered, more reliable, and less expensive than its competitors. And the Model T was uniquely suited to Ford's real innovation: the moving assembly line. First introduced in 1913 and reaching its peak at Ford's River Rouge plant in the late 1920s, the assembly line changed forever how cars—and most other products—were made in America. With the new manufacturing process, Ford's workers could build a complete car in 93 minutes.

As befits an almost mythical giant of industry, the story of Ford's childhood and rise to car-making dominance is heroic. Ford was born in 1863 in the small town of Dearborn, Michigan. His father, William, was a farmer and carpenter who had emigrated from Ireland in the 1840s, fleeing that country's devastating potato famine. His mother, Mary, died in childbirth when Ford was 13. The Ford family farm was fairly prosperous, and Henry's childhood was typical for a boy in rural America. At the age of 7, he began attending the one-room school a mile away, while also tending to chores on the farm. In his own accounts of his youth, Ford described his fascination with machinery, learning how to fix clocks and watches and even traveling from farm to farm collecting timepieces to work on. He also spent considerable time tinkering with farm machinery.

When Ford was 16, he left the farm and moved to Detroit to apprentice at a machine shop. There he learned to use specialized machine tools to make all kinds of metal goods, from valves to fire hydrants. After three years he returned to Dearborn, where he fixed and demonstrated the new steam-powered tractors that were just arriving in Michigan. In 1888 he married Clara Bryant, and they settled on a small farm where Henry also ran a sawmill. They would not stay for long; three years later, Henry was offered a job with the Edison Illuminating Company in Detroit, and the couple moved to the city.

In early 1896, Ford was talking with some other engineers at Edison. A recent issue of a magazine had described how to build a gasoline engine and

Ford's first assembly line made magnetos, a part of the Model T's ignition system. Each worker performed a specific task, over and over again.

Ford declared, "I want to build one of those." The engine featured was probably a version of the Otto engine, invented in Germany in 1876. About ten years later, two Germans, Karl Benz and Gottlieb Daimler, had independently adapted the engine to a carriage, creating the first self-propelled gasoline vehicles. By the time Ford saw the magazine story, automobiles were being made in Germany, France, and the United States.

In March 1896, Charles King, a friend of Ford's, wheeled his horseless carriage onto the streets of Detroit. Ford bicycled alongside as King accelerated to five miles per hour. But Ford was also working on his own automobile, which he completed that June. The Quadricycle had a two-cylinder gas engine, rode on four large bicycle wheels, and was steered, like a boat, with a tiller. After having to tear down a wall of his garage workshop to get it out, he, Clara, and their new son, Edsel, motored around the streets of Detroit.

Ford left the Edison company and helped found two automobile companies, both of which failed. Like many early carmakers, he also raced his own cars, and in 1901, Ford upset the speed-record-holding French champion. The fame the victory brought helped Ford start another company, and for the next several years he and his partners built ever more refined automobiles.

The Ford Motor Company was just one of thousands of small operations producing cars. Almost all of these cars were produced by hand, with skilled mechanics spending hours or days casting and altering parts and coachmakers custom-building chassis and bodies. These hand-made cars were expensive, the playthings of the rich. Ford sensed that there was a much larger market for the automobile, and by 1906 he was designing cars that could be sold at a reasonable price.

The result of this work was the Model T, unveiled in 1908. The new car had many advances—among them a cylinder block cast as one piece, an electric ignition system, and an enclosed body. Although at $825 it wasn't the cheapest car on the market, the Model T was a great value and orders flooded in. Ford sold almost 19,000 cars in 1909–1910, a figure that would jump to 78,440 during 1911–1912.

When it was fully up and running, Henry Ford's assembly line reduced the time it took to build a Model T from 14 hours to 93 minutes.

By 1924, the Ford Motor Company had sold ten million Model T's. A special car was driven from New York to San Francisco to commemorate the milestone. Here, Ford stands with that car and a model of his first, the 1896 Quadricycle.

The company could sell as many cars as it made. Assembling a car, though, was still a complicated process, requiring about 14 hours of skilled labor. More efficient manufacturing was needed. First, Ford made sure that every component made was identical, so no one had to waste time refitting or reshaping parts. He also hired Frederick Taylor, an efficiency expert, who carefully observed how people worked and suggested faster ways to do things.

Efficient mass production had been a dream of Ford's for a long time. "The way to make automobiles," he explained to a lawyer in 1903, "is to make them all alike . . . just as one pin is like another pin when it comes from a pin factory." Four years later he announced, "I will build a motor car for the great multitude. . . . It will be so low in price that no man making a good salary will be unable to own one—and enjoy with his family the blessing of hours of pleasure in God's great open spaces."

A worker typically built an entire component, like the ignition system's magneto, by himself, moving about the factory floor to collect its parts. In 1913, Ford introduced another way to do it. Inspired by the continuously moving racks of animal carcasses in a slaughterhouse, he devised a moving assembly line. A worker stayed in one place and did just one or two things to components moving by on a conveyor belt. The first assembly line was for magnetos, a component of the electrical system; one man would bolt a part on, then the component would move to the next man, who would add another part. In about 13 minutes, a complete magneto emerged.

After Ford built another moving assembly line for engines, a problem arose: The engines were being built much faster than the chassis and bodies. The logical next step was to install moving assembly lines throughout the factory. Every worker now stood in place and added a bolt or a part or something similarly small. The bodies moved along hung from chains; when done, they were lowered onto the chassis, complete with engine and interior, which were built on another line. The new process was almost 90 percent more efficient than the old way; at its peak, Ford was producing a Model T every 93 minutes.

Highly efficient production allowed Ford to cut prices, and then cut them some more, to a low of about $360 for a basic car. Although the profit margin on a single car was small, the huge volume of sales more than made up for it. Ford became one of the richest men in the world. In 1914, those profits allowed Ford to take one of his most famous actions, raising his workers' daily wages to $5, almost twice what he had been paying them. His rationale was that all of his workers should be able to afford the cars they built. Ford received great press, and many of his workers did indeed buy Ford cars.

In 1927, when the company introduced the Model A, Ford took the principles of control over production to an unprecedented level. He had bought coal mines in Kentucky, iron mines in Michigan, and a rubber plantation in Brazil in order to become independent of almost all outside suppliers. At the company's new River Rouge factory near Detroit, Ford built not only an assembly line but also blast furnaces to turn the iron and coal into steel, and a plant to turn the rubber into tires. The ideal of the self-sufficient operation that Ford may have learned as a boy on his family's homesteaded farm was finally realized in the world's biggest industrial complex.

The River Rouge plant was Ford's last great triumph. He spent his later years battling for control within the company and resisting, sometimes violently, attempts at unionization. Ford also turned to philanthropic causes. Ford built vocationally oriented schools for rural children, sponsored a homespun radio show, and built Greenfield Village, a mock small town that recreated America's agrarian and small-industrial past. He died in 1947, two years after his grandson, Henry Ford II, assumed control of the company. Today, Ford is still run by the family; William Clay Ford, Jr., was named chairman in 1999 and has since been credited with making the company "greener."

ROBERT GODDARD

On October 19, 1899, Robert Goddard climbed a cherry tree in his yard and had a vision that would define his career and eventually send men to the moon. A year after being enthralled by H. G. Wells's interplanetary science-fiction tale *The War of the Worlds*, the 17-year-old Goddard imagined traveling through space himself. "As I looked to the fields at the east," he later wrote, "I imagined how wonderful it would be to make some device which had even the *possibility* of ascending to Mars."

Although Goddard's idea that day of how get to Mars was wrong—it involved using centrifugal force—the epiphany changed him. "I was a different boy when I descended the tree," he continued. "Existence at last seemed very purposive." Goddard pursued that vision for the next 45 years, building and launching rockets and devising the basic concepts of rocketry and space flight—concepts used in all succeeding rockets, from Germany's V-2 missiles of World War II, to the Saturn V that took men to the moon, to the boosters that launch satellites today.

Although his ascent into the cherry tree was a turning point in Goddard's career, it was far from his first thought of flight. Born in 1882 to parents whose roots in America went back to Puritan settlers, Robert Goddard grew up in a house where scientific and technological progress was always prized. His father, Nahum Goddard, was an occasional inventor and early adopter of late-nineteenth-century advances such as electricity. Goddard himself began experimenting at an early age; when he was just five, he tried using static electricity and zinc from a battery to enable him to jump very high, and in high school he attempted to melt aluminum to create a balloon. (Neither experiment worked.)

Goddard's investigations became more focused, and more successful, in college. Beginning as an undergraduate at the Worcester Polytechnic Institute, Goddard worked to understand the challenges of flight, especially rocket flight. He continued his education at Clark University, where he earned his master's in 1910 and his doctorate in 1912. There were two main problems facing him: the general belief that nothing could be propelled in the vacuum of

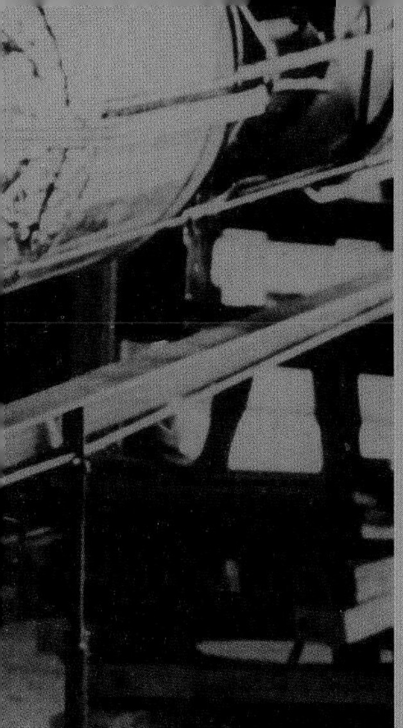

In December 1925, Robert Goddard fired the first liquid-fueled rocket that could lift its own weight—it moved about an inch. Four months later, he traveled to his aunt's farm in Auburn, Massachusetts, to set up a bigger rocket.

That rocket, seen here on its launch frame, emitted a flame and a steady roar and then began to rise, eventually reaching a height of 41 feet during its two-and-a-half-second flight.

The press- and publicity-shy Goddard moved his research to Roswell, New Mexico, in 1930. His wife, Esther, filmed this launch of a rocket code-named Nell in August 1937.

space, and the limited power of black powder, the fuel then used for projectiles. In 1907 he proved mathematically that a rocket could work in a vacuum—Newton's third law of thermodynamics, "For every action there is an equal and opposite reaction," still applied. And in 1909 Goddard hit upon a solution to the problem of fuel. Liquid fuels, like liquid hydrogen and oxygen, could increase a rocket's power tenfold or more. Unfortunately, neither fuel was available at the time. (Unknown to Goddard, a Russian named Konstantin Tsiolkovsky, also keenly interested in space flight, had come up with a similar idea a few years earlier.)

In 1914, after recovering from tuberculosis, Goddard became a professor of physics at Clark University. That year he also received two patents, one for a rocket using liquid fuel, the other for a multistage rocket. The next year, he set up a vacuum chamber in his lab and ignited a rocket in it, proving empirically that an engine could work in the absence of air.

In the mid-1910s Goddard first confronted one of the major non-technical challenges that would plague much of his career: money. Rocketry was an expensive pursuit, and his experiments soon drained his resources. In 1917, he finally received a small grant ($5,000) from the Smithsonian Institution to continue his studies; the funds would continue through 1929. In 1919, he submitted a report detailing his expenditures and outlining his progress. The report also suggested, probably in jest, that one way to prove his theories was to send a rocket equipped with flash powder to the moon. When humans saw the flash, they'd know it worked. The report, "A Method of Reaching Extreme Altitudes," was published in late 1919, and reached the press in early 1920.

Goddard was unprepared for the reaction to his seemingly obscure paper. Journalists latched on to his idea to send a rocket to the moon, dubbing the scientist "Moon Man" and running headlines such as "Modern Jules Verne Invents Rocket to Reach Moon." The *New York Times* ran a skeptical editorial that attacked him personally: "Professor Goddard does not know the relation of action to reaction, and of the need to have something better than a vacuum against which to react. . . . Of course he only seems to lack the knowledge ladled out daily in high schools."

For the rest of his career, Goddard was publicity-shy, avoiding the press and working in out-of-the-way places. But he continued to work. Goddard spent the early 1920s designing and refining rocket engines that used liquid fuel. In December 1925, he fired a rocket that could lift its own weight—a crucial test. It moved only about an inch, but, Goddard wrote, "it shows that a larger rocket constructed on the same plan could raise itself to considerable altitudes."

Four months later, on March 16, 1926, he journeyed to his aunt's farm in Auburn, Massachusetts. There he set up a larger liquid-fuel rocket, a thin, spindly looking device about ten feet tall. At the bottom of the rocket were tanks holding gasoline and liquid oxygen. Fuel lines ran to the top of the rocket frame, to a combustion chamber. The two fuels combined in the chamber and were ignited by a flame; the exhaust, expelled through a nozzle, provided thrust. That afternoon he lit the rocket; for a few seconds the engine just roared, then it slowly lifted off the ground. The world's first liquid-fueled rocket flew for about two and a half seconds and achieved a maximum height of 41 feet. In his notes, Goddard wrote, "It looked almost magical as it rose . . . as if it said: 'I've been here long enough. I think I'll be going somewhere else, if you don't mind.'"

It was a humble beginning for space flight, but more experiments, both successes and failures, followed. In 1930, Goddard moved his research to Roswell, New Mexico, to avoid prying eyes. There he devoted himself full-time to refining his rockets and soon introduced such innovations as gyroscopic stabilization and fins in the path of the rocket's blast to keep it stable. The rockets grew larger and faster, too. In 1935, one flew faster than the speed of sound, and two years later a rocket reached 9,000 feet, the highest launch of Goddard's career.

Goddard always believed that his rockets had important military uses. During World War I, he worked with the army to develop small rockets and launchers, innovations that eventually produced the bazooka. As World War II approached, he became worried about German rocket research; he knew German scientists had been eager followers of his work. When he approached the army about using his rockets as weapons, though, he was turned away. They did not see any practical potential in his inventions. Instead, Goddard was put to work designing jets to assist planes as they took off.

The German military, on the other hand, saw plenty of potential in rocketry. Led by Wernher von Braun, German scientists used Goddard's patents to design the V-2 rocket, a powerful weapon with a range of up to 200 miles. Flying extremely fast and at a great altitude, the V-2 was impossible to defend against.

KONSTANTIN TSIOLKOVSKY—THE FATHER OF SPACE TRAVEL

Although Robert Goddard was a rocketry visionary, sending the world's first liquid-fueled rockets more than a mile into the sky, many of his ideas were first thought of by Konstantin Tsiolkovsky, a deaf schoolteacher working in provincial Russia. But while Tsiolkovsky developed some of these concepts first, he was interested not in rocket propulsion itself, but where it could take people. From the time he was a teenager, he dreamed of space flight and man's eventual life in space.

Born in Ijevskoe, Russia, in 1857, Tsiolkovsky was one of 18 children in a modest farm family. At the age of 10 he lost his hearing as a result of scarlet fever, which prevented him from finishing school. He continued to read on his own, though, and in 1880 he became a math and physics teacher in a public school while continuing his investigations into both science and philosophy.

From the 1880s until his death in 1935, Tsiolkovsky published dozens of papers and books about his scientific ideas. His 1897 paper, "Exploration of Outer Space with Reactive Devices," laid out the basic idea of humans using rockets to travel through space. Six years later, he wrote a paper outlining spacecraft that used the reaction of liquid hydrogen and liquid oxygen for thrust—several years before Goddard made the same discovery.

Tsiolkovsky's work was always theoretical—he never built any of the rockets or ships he discussed. But his theories were far ahead of his time. In one paper, he drew a spacecraft that looks

The first rockets were launched against Paris and London in September 1944; before the war ended, about 1,100 V-2s were used, killing some 2,800 people.

Robert Goddard died in August 1945, just days before the end of the war. His work was continued by the army, however, which had finally seen the promise of rockets. In the years following the war, they launched 67 captured V-2s and began developing a new generation of missiles. The V-2 tests led to the Redstone and Jupiter rockets, which in turn led to the massive, multistage Saturn rocket that powered America's Apollo launches.

Recognition for Goddard's pioneering achievements did come, but only long after he died. Today he is widely regarded as the father of rocketry, and NASA's first space flight research facility, the Goddard Space Flight Center, located in Greenbelt, Maryland, was named after him. Even the *New York Times* eventually came around. On July 17, 1969, three days before Neil Armstrong and Buzz Aldrin walked on the moon, the paper's editorial page noted "it is now definitely established that a rocket can function in a vacuum as well as in an atmosphere. The *Times* regrets the error."

A late-model rocket without its casing sits on its assembly frame in Roswell in 1940. Goddard, at far left, inspects it with three of his colleagues: Nils Ljungquist, Albert Kisk, and Charles Mansur.

remarkably modern: An astronaut sits at the front of a long rocket filled with hydrogen and oxygen. He conceived of steering vanes much like Goddard's later versions, pressurized cabins and space suits, gyroscopic stabilization, and air locks to permit entering and leaving a ship. He also had a vision for permanent human outposts in space, including space stations that spun to simulate gravity and closed biological systems that could provide food and oxygen.

After the Russian Revolution in 1917, other scientists began to notice and pursue some of his ideas. When the Soviet space program was in its heyday, from the launch of Sputnik in 1957 through the Soyuz missions of the 1970s, Tsiolkovsky was a national hero. It is still remarkable how many of his ideas and predictions have become reality—the U.S. space shuttles, for example, rely on essentially the same kind of engine he sketched out in 1903, and the International Space Station has confirmed his early vision of a habitat continuously orbiting the Earth.

Paul MacCready

Anyone who happened to be on the English Channel on the morning of June 12, 1979, must have done a double take. Moving slowly through the air, just a few feet above the choppy waters, was one of the oddest looking flying machines ever built. Almost 100 feet wide and sheathed in a shiny, semitransparent skin, the craft, called the *Gossamer Albatross*, had a big propeller at its back but no engine to speak of. Instead, in an enclosed pod hanging beneath its huge wing, was Bryan Allen, a bicycle racer, furiously pedaling to make the improbable craft go. After almost three hours of physical effort, and covering more than 22 miles, Allen gently landed the plane on the beaches of England. The *Albatross*'s inventor, Paul MacCready, who had been in a boat following the slow, tense journey from the shores of France, rushed to join the celebration.

Men have dreamed of flying under their own power for thousands of years. A Greek myth tells of Icarus, whose father Daedalus built wings of feathers and wax so the two of them could fly like birds. The dream was revived in the Renaissance, when Leonardo da Vinci sketched complicated machines that would allow men to fly. After the Wright Brothers built their first airplane in 1903, the search for a way to fly without a motor entered a new phase. Throughout the twentieth century, experimenters built a succession of human-powered planes, most of which never left the ground.

Although he didn't begin building a human-powered airplane until he was 51 years old, Paul MacCready had been involved with flight for most of his life. As a boy growing up in New Haven, Connecticut, he was fascinated with butterflies and moths, and his interests soon included model airplanes, too. His first model plane came from a kit, but after that he was building his own, from balsa wood and microfilm and glue. He didn't build only standard aircraft. "For some reason I got interested in a variety of things," he says. "Ornithopters, autogyros, helicopters, indoor models, outdoor models. Nobody seemed to be quite as motivated for the new and strange as I was."

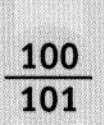

Paul MacCready's second human-powered airplane, the *Gossamer Albatross*, flew across the English Channel in June 1979, pedaled and piloted by Bryan Allen.

After the success of the *Albatross*, MacCready began to work on the challenge of solar-powered flight. His first solar plane was the *Gossamer Penguin*, here on a test flight at NASA's Dryden Flight Research Center.

The *Pathfinder*, a lightweight, solar-powered, remotely piloted flying wing made its first flight in 1994. Though it flew at only 15 to 20 mph, it could stay in the air for great lengths of time and in 1997 reached an altitude of 71,500 feet.

FLYING A DINOSAUR

Paul MacCready has made a career of flying things that shouldn't fly. The aviation challenge he took up in 1984, though, was very different from the flights of the *Gossamer Albatross* and the *Pathfinder*. Rather than exploring a new way to fly, he and his team at AeroVironment set out to resurrect a 65-million-year-old flyer, the pterodactyl.

Pterosaurs, or flying dinosaurs, were the first large creatures to fly and may be the ancestors of today's birds. In 1972, paleontologists in Texas discovered the well-preserved fossils of the largest pterosaur known, *Quetzalcoatlus northropi*, which had a wingspan of 36 feet. In 1980, MacCready heard about the fossil, went to see it, and came away amazed that it could fly at all; its large size pushed its flight to the edge of physical possibility. By 1983, after more study, he became convinced that he could build an air-worthy model of *Quetzalcoatlus*.

The money to build such a flying dinosaur came in 1984, in the form of a commission from the Smithsonian's National Air and Space Museum, which was preparing to make an IMAX movie, *On the Wing*, relating manned and natural flight. With sponsorship from the Smithsonian and the Johnson Wax Company, MacCready got to work.

Figuring out how a pterodactyl flew was a difficult task. The dinosaurs had no tails, so how they steered and maintained stability was a bit of a mystery. (It was "like trying to fly an arrow with the tail feathers in front," according to the director of the project.) Making a model's wings flap realistically and with enough power to lift it was an even bigger challenge. Technology and intensive aerodynamic study came to the rescue. MacCready's team used lightweight carbon-fiber tubes to build a skeleton, small motors to control its wings, and a sophisticated computer

After winning several model plane competitions, MacCready graduated to sailplanes, gliders built to soar and maneuver in the wind. "It's a very scientific sort of hobby," he recently told an interviewer. "It's not just like going out and rowing a boat. You get involved in the science of the vehicle, because the vehicle has to be efficient." When he was 23, he won the U.S. National Soaring Championship.

During World War II MacCready trained to be a navy pilot, but he never flew in combat. After the war, he returned to his education, though, he recognizes now, he struggled with a mild form of dyslexia. In 1947 he earned a degree in physics at Yale, then went on to receive his Ph.D. in aeronautics from the California Institute of Technology in 1952. Although flying was still one of his great passions, MacCready didn't want to join a big aeronautics or engineering firm. "There were already enough brilliantly qualified people working in aerospace," he says, "and the typical projects were big things where they'd hire another acre of engineers and they'd work on something where in seven years it would fly. It's kind of hard to get motivated." In a smaller operation, he adds, "a person can make a much greater contribution to society." After CalTech he cofounded Meteorology Research, a company that pioneered weather modification, including cloud seeding. In 1970 he left to start another company, AeroVironment, to focus on new energy sources such as solar and wind power.

It was bad news that got MacCready working on human-powered flight. He had guaranteed a loan for a relative's business. It failed, and he was stuck with a $100,000 debt. Daydreaming one day in 1976, he recalled that there was a cash prize for a successful human-powered flight: the Kremer Prize, with an award of £50,000. With the British pound worth about two dollars at the time, he remembers, "the Kremer Prize, in which I'd had no interest, was just about equal to my debt. Suddenly human-powered flight seemed important."

MacCready began thinking of ways to build a plane that could win the prize, which required flying around a figure-eight course. (Other human-powered planes had flown, but they couldn't make turns.) The Aha! moment, as he calls it, came on a family vacation. On the side of the road, MacCready watched hawks and vultures in flight, calculating their flight speed and turning radius. He began thinking about how scaling—making something bigger or smaller—affects a wing's aerodynamic lift.

In 1986, MacCready built and flew a lifelike model of a pterodactyl, complete with beak, 18-foot wingspan, and a light coat of fur.

to control their flapping. The model pterodactyl's head and bill acted as a kind of rudder, and its wings were moved forward and back to give it lift and stability.

Their finished model, with an 18-foot wingspan and sporting fur and a realistic paint job, was tested in a wind tunnel and also mounted to the roof of a speeding van. In January 1986, after almost two years of work, experiment, and crashes, MacCready's pterosaur was ready to fly. With the flight team and a crew of IMAX filmmakers standing by, *Quetzalcoatlus* was launched into the blue sky above Death Valley, California. Soaring 500 feet into the air, the relic of the past gently glided through graceful turns above the craggy landscape, its re-created wings gently flapping and its long-billed head turning in the direction of its flight. Nothing, it seemed, is too ungainly or unlikely for MacCready to fly.

The *Helios*, another solar-powered flying wing, has a wingspan of 247 feet. It is meant to reach extremely high altitudes and stay aloft for months at a time.

"For some reason I got interested in a variety of things," he says. "Ornithopters, autogyros, helicopters, indoor models, outdoor models. Nobody seemed to be quite as motivated for the new and strange as I was."

He realized that as a wing is built bigger, it requires less power to keep it aloft. If built a lot bigger, it requires a lot less power. A very light, 96-foot-long wing (as big as a DC-9's) built like a hang glider's wing would only require about 0.4 horsepower to make it fly—about the power a good bicyclist can produce.

"I had the advantage of no background in aircraft structures," MacCready says. "So rather than being lured into doing airplane wings in standard ways, I just went back to fundamentals. How do you make a 96-foot wing that weighs practically nothing?" It took six months of sometimes nonstop work for MacCready and his team, made up of friends, colleagues, and family, to build the *Gossamer Condor*. They used simple materials that were light and could be fixed easily, including Mylar, piano wire, aluminum tubing, and lots of tape. When it was finished, the *Condor* weighed just 70 pounds, didn't look much like a plane, and didn't always act like one. (Bryan Allen, the pilot/engine, said flying it was "like pedaling a house.") On August 23, 1977, with Allen at the pedals, the *Condor* flew the Kremer course. Today, the *Gossamer Condor* hangs in the Smithsonian Institution's National Air and Space Museum, alongside the Wright Brothers' plane, the *Spirit of St. Louis,* and an Apollo Lander.

There was another prize to be won: The Kremer Prize's sponsor put up £100,000 for the first human-powered aircraft to cross the English Channel. With sponsorship from Dupont, which made his planes' Mylar skin, MacCready set about building a second plane, the *Gossamer Albatross*. Two years later the human-powered *Albatross* crossed the Channel.

From human power MacCready moved on to another challenge, harnessing the sun's energy. Building on the design and success of the first Gossamer planes, he constructed the *Gossamer Penguin*, which in 1980 became the world's first solar-powered airplane. The craft caught the eye of the Department of Defense, which wanted something that could stay up in the air for long periods of time. Eventually, the project was taken up by NASA, which was interested in creating an airborne platform for atmospheric observations. The unmanned *Pathfinder* soared to a height of 71,530 feet in 1997; its successor, the *Pathfinder Plus*, reached 80,201 feet in 1998. The next step in this evolution is the 200-foot-wide *Helios*, which might reach 100,000 feet and stay aloft for months at a time. MacCready envisions a fleet of these craft in the air permanently, serving as a high-performance communications platform.

In the late 1980s, MacCready applied his knowledge of solar power and aerodynamics at the ground level. Working with General Motors, he designed a solar-powered car, the Sunraycer, for the World Solar Challenge race across Australia. Powered by 8,800 photovoltaic cells that covered its super-aerodynamic body, the Sunraycer dominated the 1,867-mile race, winning by two full days and averaging 41 miles per hour. Although the light, low-slung car could never be called practical transportation, its success spurred GM's effort to create the Impact, the auto giant's first electric vehicle. The popularity of that prototype led to GM's EV-1, a production electric vehicle that can go from 0 to 60 mph in eight seconds.

MacCready is interested once again in model planes. For the past few years, he's been working on small aircraft with built-in cameras to send images back to the ground. "You are looking out as if you were a little creature inside," he explains. "You can soar with the birds. It's just like you're one of them." This new vehicle fits in perfectly with MacCready's interest in finding a balance between nature and technology. "The overall goal is a sustainable world," he says. "Not consuming nonreplenishable resources, not getting steadily more dependent on foreign oil, and not causing global climate change." How can a toylike plane that lets you fly with birds help reach those goals? "If I hadn't been doing the ornithopters 60 years ago there wouldn't have been a *Gossamer Condor* or *Albatross* or Impact car or a mandate in California on zero emission vehicles. It's a toy, but it's a pretty important toy."

The Sunraycer, a solar-powered car designed by MacCready and General Motors, won the first World Solar Challenge Race in 1987, averaging 41 mph over 1,867 miles.

GARRETT MORGAN

Most inventors wait for years or even their entire life for the chance to prove the value of their work. Not Garrett Morgan. On July 25, 1916, news came to the Cleveland businessman that disaster had struck—and Morgan had his chance. Thirty-two men building a tunnel beneath Lake Erie had been trapped by an explosion. The tunnel was full of debris, smoke, and toxic gases. Rescue parties had been sent in but had not returned. Morgan and his brother Frank rushed to the scene, equipped with a device he had invented four years earlier, the Safety Hood, a version of what is now called a gas mask.

His invention consisted of a large, heavy canvas hood fitted with small glass windows to see through. A long tube ran from the bottom of the hood to a pouch containing a sponge that could filter smoke, ammonia, and other dangerous substances from incoming air. (Morgan's inspiration for the hood and its long tube was seeing a circus elephant use its trunk to reach fresh air from inside a stifling hot big top.) The hood was big enough to hold enough air to breathe for at least 15 minutes, even when the filter was plugged up to keep dense smoke out.

Morgan and his brother donned Safety Hoods and entered the smoke-filled tunnel. After an agonizing wait, the two emerged, each carrying a worker on his back. Two other men joined them, and they went back into the tunnel for other survivors. Although many of the workers could not be saved, Morgan's rescue made him a hero in Cleveland, and his Safety Hood garnered national attention.

The Safety Hood was already known to fire departments around the country. In 1914 Morgan and his invention had won a gold medal at the Second International Exposition of Safety and Sanitation in New York, and testimonials of its usefulness had come from as far away as New Haven, Connecticut. But the wave of publicity following the Cleveland rescue was not necessarily a good thing. Garrett Morgan was black, and when white buyers of his hoods learned this, many canceled their orders. Morgan had known that the color of his skin might prevent people from buying his products; when he demonstrated the Safety Hood in New Orleans in 1914, he did so under an assumed name, "Big Chief" Mason, and claimed to be a "full-blooded Indian, from the Walpole Reservation in Canada." Although he had received a patent for the hood, he did not dare take public credit for the invention. However, when the Germans threatened to use chemical weapons during World War I, Morgan gave his patent to the U.S. government, which developed a gas mask based on his work.

Garrett Morgan was born in 1877 in Paris, Kentucky, to Sydney and Elizabeth Morgan, both former slaves. As a child, he attended a local school and worked on his parents' farm. Morgan dreamed of furthering his education, but in rural Kentucky there was little opportunity to do anything but farm. At the age of 14, he left home and headed north, working for a while in Cincinnati as a handyman, then moving to Cleveland in 1895. There he found work fixing sewing machines for a clothing manufacturer. Morgan worked for several sewing machine makers until 1907, when he had saved enough money to open his own shop selling and repairing sewing machines. It was the first of several businesses he would start.

Morgan's first big invention came two years later, after he had expanded his business to include a tailor's shop. (His wife was an accomplished seamstress.) He had noticed that sewing machine needles moved so fast that they sometimes scorched fabric. Morgan began to experiment with a variety of liquids that could polish a needle so it would sew smoothly and without friction. As he left his workshop one night after working with one such liquid, he wiped has hands dry on a piece of pony fur cloth. When Morgan returned, the fur had become straight and was sticking up. Intrigued, he applied the liquid to the wiry fur of a neighbor's dog. It, too, became straight. Next, Morgan tested it out on his own hair. It worked on humans, too, and Morgan soon christened the product G. A. Morgan Hair Refining Cream.

The G. A. Morgan Hair Refining Company was a big success. Morgan added several other products to its line, including hair dyes for men and women and a "hair-growing" cream. The company sold its goods by mail order and through a network of salespeople. With this business model, Morgan may have been emulating the success of Madame C. J. Walker, who had started a similar business a few years earlier in St. Louis. Walker and a small army of "agents" demonstrated and sold her hair straighteners and other beauty products door-to-door. Walker's products quickly became famous, and by the 1910s Madame C. J. Walker was a millionaire, one of the first African Americans to achieve such financial success.

While Morgan's hair-straightening cream didn't reach the level of success of Madame Walker's, it did make him one of Cleveland's most successful African American citizens. By the early 1920s, he was the proud owner of a new automobile, at a time when cars were still something of a novelty. The streets of Cleveland, as was the case in most American cities, were filled with horses, horse-drawn carriages, bicycles, pedestrians, and cars. One day in 1922, Morgan was driving his two young sons around town when they witnessed a horrible accident between a car and a horse-drawn carriage.

The shock of witnessing this accident quickly turned into a resolve to do something to prevent similar occurences. Efforts had already been made to control the flow of traffic. The most common solution was using a police officer to direct traffic, although that required a lot of labor and the officer was not visible over the roofs of cars. Some people had designed signals, but those usually had only "stop" and "go" settings. While these were useful, the inventor's granddaughter, Sandra Morgan, explains, "no one would know that the traffic pattern was about to change," which could lead to accidents.

Morgan's solution was the folding traffic signal. Mounted above traffic on a post, the signal had folding arms with "stop" and "go" written on their various sides. The arms folded into four different positions. When it was time for east-west traffic to move, for example, a traffic attendant would turn a crank at the

MADAME C. J. WALKER

Garrett Morgan's initial fortune came from selling hair-care products made specifically for the African American market. But while his work on the gas mask and traffic signal, among other things, was original, his cosmetic empire followed a path forged a few years before by Madame C. J. Walker, who may have been America's first black female millionaire.

Born Sarah Breedlove in rural Louisiana in 1867, Madame C. J. Walker—she adopted the last name and title during a brief marriage—worked as a washerwoman until 1905. That year, Walker claimed, she had a dream in which a treatment for African American women's hair was revealed to her. This treatment comprised a shampoo, a "hair grower," and a method for straightening hair. (Previously, many African American women straightened their hair with a flatiron.)

Madame Walker began selling her hair products in St. Louis in 1905, but she soon moved to Denver and gradually increased her business. One of the keys to Walker's success was her business model, which relied on door-to-door sales by a network of "Walker agents" that eventually spanned the entire country and parts of the Caribbean. This approach was successful for several reasons; for one, it allowed her to reach a commercially disenfranchised market, African American women. Face-to-face contact and personal demonstrations were also powerful enticements to buy her products. Many Walker customers also became agents, which helped business increase exponentially.

By the late 1910s, the Madame C. J. Walker Manufacturing Company had some 2,000 agents actively selling; her clients included celebrities such as Josephine Baker. Madame Walker's own fortune had increased beyond $1 million. Before she died in 1919, she made generous gifts to several philanthropic enterprises, including schools for African American girls. After Madame Walker's death, her daughter, A'Lelia, continued some of her mother's social activities, notably by holding an artists' and writers' salon in Walker's luxurious mansion in New York's Harlem. This salon, in part, helped kick-start the Harlem Renaissance, which flourished through the 1920s.

Garrett Morgan's first successful invention was a hair-straightening cream—he called it a "hair refiner"—which he sold by mail order and through a network of salespeople.

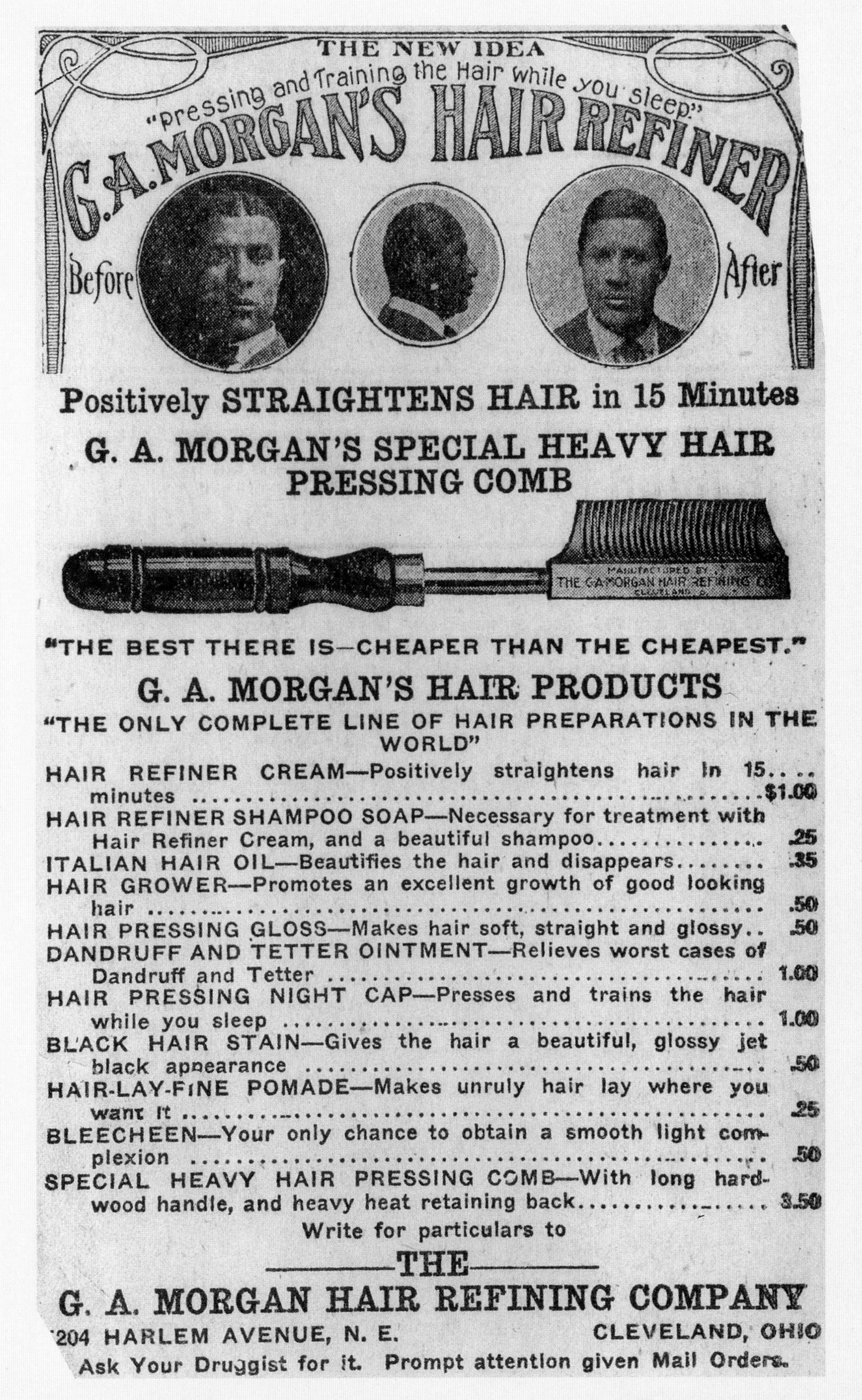

THE NEW IDEA

"Pressing and Training the Hair while you sleep."

G. A. MORGAN'S HAIR REFINER

Before After

Positively STRAIGHTENS HAIR in 15 Minutes

G. A. MORGAN'S SPECIAL HEAVY HAIR PRESSING COMB

"THE BEST THERE IS—CHEAPER THAN THE CHEAPEST."

G. A. MORGAN'S HAIR PRODUCTS

"THE ONLY COMPLETE LINE OF HAIR PREPARATIONS IN THE WORLD"

HAIR REFINER CREAM—Positively straightens hair in 15 minutes $1.00
HAIR REFINER SHAMPOO SOAP—Necessary for treatment with Hair Refiner Cream, and a beautiful shampoo .... .25
ITALIAN HAIR OIL—Beautifies the hair and disappears .... .35
HAIR GROWER—Promotes an excellent growth of good looking hair .... .50
HAIR PRESSING GLOSS—Makes hair soft, straight and glossy.. .50
DANDRUFF AND TETTER OINTMENT—Relieves worst cases of Dandruff and Tetter 1.00
HAIR PRESSING NIGHT CAP—Presses and trains the hair while you sleep 1.00
BLACK HAIR STAIN—Gives the hair a beautiful, glossy jet black appearance .... .50
HAIR-LAY-FINE POMADE—Makes unruly hair lay where you want it .... .25
BLEECHEEN—Your only chance to obtain a smooth light complexion .... .50
SPECIAL HEAVY HAIR PRESSING COMB—With long hardwood handle, and heavy heat retaining back 3.50

Write for particulars to

—THE—

G. A. MORGAN HAIR REFINING COMPANY

204 HARLEM AVENUE, N. E. CLEVELAND, OHIO

Ask Your Druggist for it. Prompt attention given Mail Orders.

In 1912, Morgan invented the "safety hood," a forerunner of the gas mask. The hood completely covered a firefighter's head, to protect him from smoke and gases. Air was filtered through the hood's long "tail."

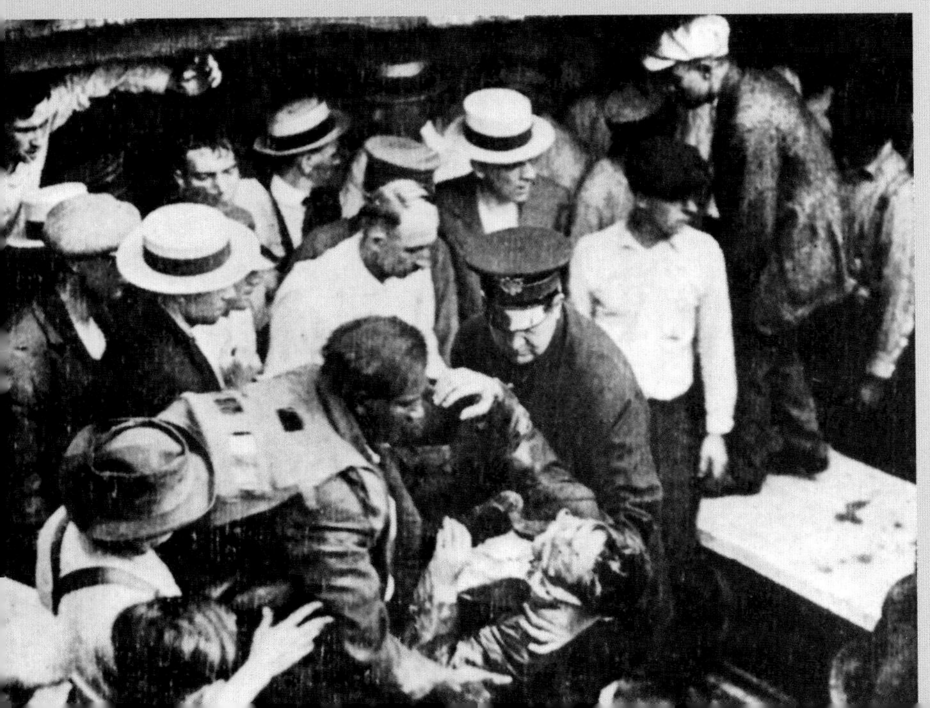

In July 1916, Morgan's safety hood was put to its greatest test, in a rescue of workers trapped in a mine beneath Lake Erie. Morgan and his brother donned hoods and rushed in to help; his action made him a hero.

base of the post and two signs saying "go" would be revealed. When north-south traffic needed to move, the folding signal would be rotated 90 degrees, and now three bold signs reading "stop" would face the east-west traffic. If pedestrians needed to cross, or traffic needed to be stopped for any reason, the sign could fold up to read "stop" in all four directions. Finally, the arms of the signal could be raised halfway between "go" and "stop," to indicate that motorists should proceed with caution. This was the ancestor of the yellow light in the middle of the modern traffic signal. A patent for the folding signal was issued in 1923, and Morgan soon sold the rights to General Electric for $40,000.

With the success of his hair products, the traffic signal, and his tailoring shop, among other ventures, Morgan had become a prominent and wealthy man. But he never forgot—or let others forget—that he had been born poor and had struggled against discrimination. "His big line was, 'I have not been well educated, but I am a graduate from the school of hard knocks and cruel treatment,'" recalls Sandra Morgan. And he always encouraged his children and grandchildren to pursue their education, insisting, "Work with your head, not with your hands."

Morgan's efforts at education and advancement were also directed at Cleveland's African American community. In 1920 he helped found a weekly newspaper, the *Cleveland Call*; in 1925 it became the *Call & Post*, and it is still published in several Ohio cities. In 1931 Morgan ran for a position on the Cleveland city council, and while he did not win, his candidacy established the black community as an important constituency. Although he faced several setbacks—the depression hit his business very hard, and in 1943 he lost most of his sight—he continued to both invent and work as a community leader until his death in 1963.

In 1922, after Morgan witnessed a horrible traffic accident, he turned his inventive mind to automobile safety. A year later, he was awarded the first patent for a traffic signal.

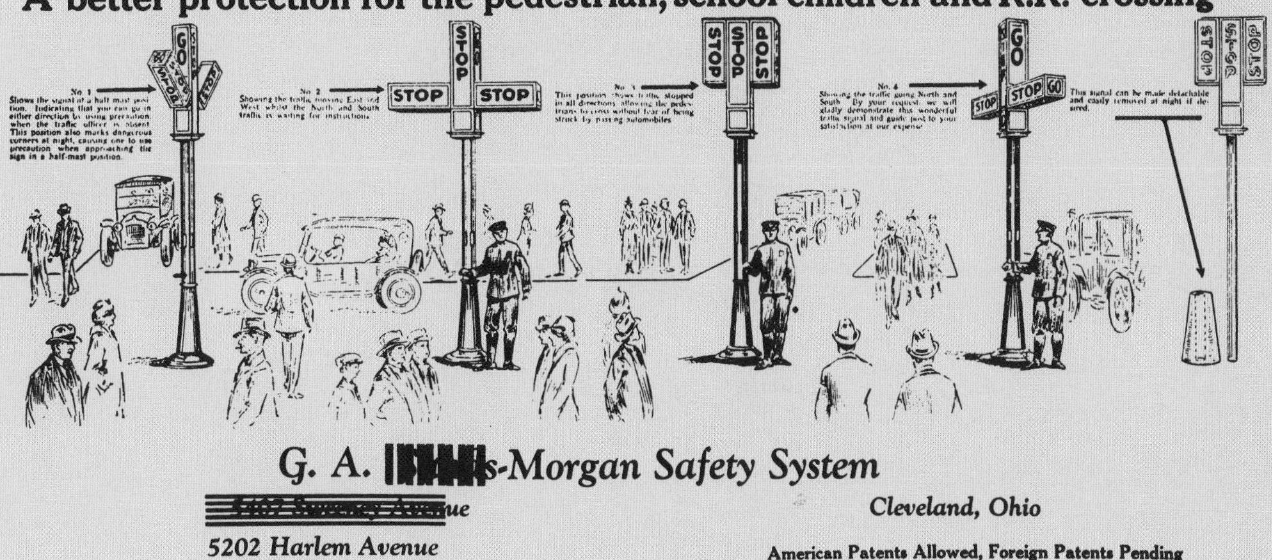

ELMER SPERRY

Sailors have been using magnetic compasses to find their way since the twelfth century, first with natural lodestone and later with magnetized needles. In the late nineteenth century, however, the invention of metal-hulled ships threatened this navigational necessity: All of that metal surrounding a compass made it erratic. The problem was worse in the submarines just being developed. And once the airplane was invented, it became clear that aerial navigation would pose an even more severe problem.

Around 1907, the engineer and lifelong inventor Elmer Sperry became interested in gyroscopes, which until then had been treated as a novelty. A gyroscope always spins in the direction of its original motion, until friction slows it down enough to make it unstable. Sperry used this property to create a gyroscopic compass that would always "point" north. He also used heavy gyroscopes to help ships resist rolling on high seas. His gyrocompass and gyrostabilizer were installed in large navy vessels and soon revolutionized marine navigation.

Sperry was born in the small town of Cortland, New York, in 1860. His mother died soon after his birth, and because his father had to earn a living—variously as a lumberman, a carpenter, a wagonmaker, and the owner of a traveling carnival—Elmer was raised by his aunt, Helen, on a farm outside Cortland. Like many other inventors of the late nineteenth and early twentieth centuries, Sperry acquired his mechanical education from farming—a small farmer needed to know his machines inside and out, to fix them and to improve them. Sperry's interests soon went beyond farm equipment; as a boy, he built model windmills and water wheels. At the age of six, he gave his aunt what may have been his first invention, a hand mill for grating horseradish.

In 1907, Sperry began to experiment with gyroscopes—especially their potential to stabilize automobiles. Within a couple of years, he had turned his attention to stabilizing ships; this 30-ton gyro wheel would eventually be installed in a 10,000-ton vessel.

When he was about 10 years old, Sperry moved into Cortland with his grandparents. Although the town was small, it was an important stop on the railroad and boasted several industries. While he was a teenager, Sperry had the chance to observe the goings-on at the blacksmith's shop, the machine shop, the rail yard, and elsewhere. He also began working at some of these businesses, including the local foundry and the book bindery. When Sperry was 16, he helped build an electric dynamo at the foundry, and he traveled to the Centennial Exposition in Philadelphia where it was exhibited. The exposition was a great place for him to see the fruits of modern technology, including the Jacquard loom, which was controlled by punch cards.

After using gyroscopes to stabilize ships and compasses, Sperry turned to aviation and built a "gyropilot" from small gyroscopes. Here, the famous aviatrix Amelia Earhart poses with a 1930s version.

In Cortland, Sperry attended the local college for about two years, studying science and engineering, but he never got his degree. By the time he left the school in January 1880, however, his main interest had become clear: electricity. The announcement of Thomas Edison's incandescent light bulb in 1879 had excited him along with much of the country. That year, he had built his own dynamo and a new kind of arc lamp, and in 1880 he moved to Chicago to begin manufacturing them. He had high hopes; in 1883 he wrote, "Do you suppose for one moment that I am going to fail in great Chicago where one can turn a hundred ways—no sir."

The field of electricity and lighting was wide open when Sperry arrived in Chicago. For several years, his company was a success, selling lighting systems to cities and towns throughout the Midwest; his lights even adorned the tower of the Chicago Board of Trade. He was a better engineer than a businessman, though, and several of his companies struggled as technologies changed. From 1880 to 1910, Sperry was involved in a series of industries, including lighting, engines, rail cars, batteries, automobiles, and chemistry.

In 1907, for reasons that are not entirely clear, Sperry turned his attention to the gyroscope. The French physicist Jean-Bernard-Léon Foucault had built the first proper gyroscope in the 1850s: a heavy spinning wheel mounted in a frame that let it rotate in any direction. The gyroscope's most useful properties are its stability—the wheel will keep spinning in its initial direction—and its precession, the tendency to move at a right angle to any force exerted on it. One of the problems of capitalizing on these characteristics was that eventually friction would slow the wheel down. In 1890, an engineer solved this by building an electric motor to drive the wheel.

Sperry had been building automobiles for several years and thought to use a spinning gyroscope to stabilize a car. The theory was simple: If a car with a large horizontally spinning gyroscope (it was mounted beneath the seats) tipped, the angular force exerted by precession would resist the tip. He patented the idea, but could find no interest among carmakers.

Despite the rejection, Sperry still thought the idea was a good one, and he applied it next to ships. During an 1898 voyage to Europe, he had experienced an unpleasant effect of a rolling ship—seasickness. Using the same basic concept as his auto stabilizer, Sperry built a large, heavy gyroscope meant to be

Elmer Sperry's early career was devoted to the burgeoning business of electricity. In April 1884, he helped found the American Institute of Electrical Engineers, now the IEEE; at Sperry's right is Nathaniel S. Keith, secretary of the group.

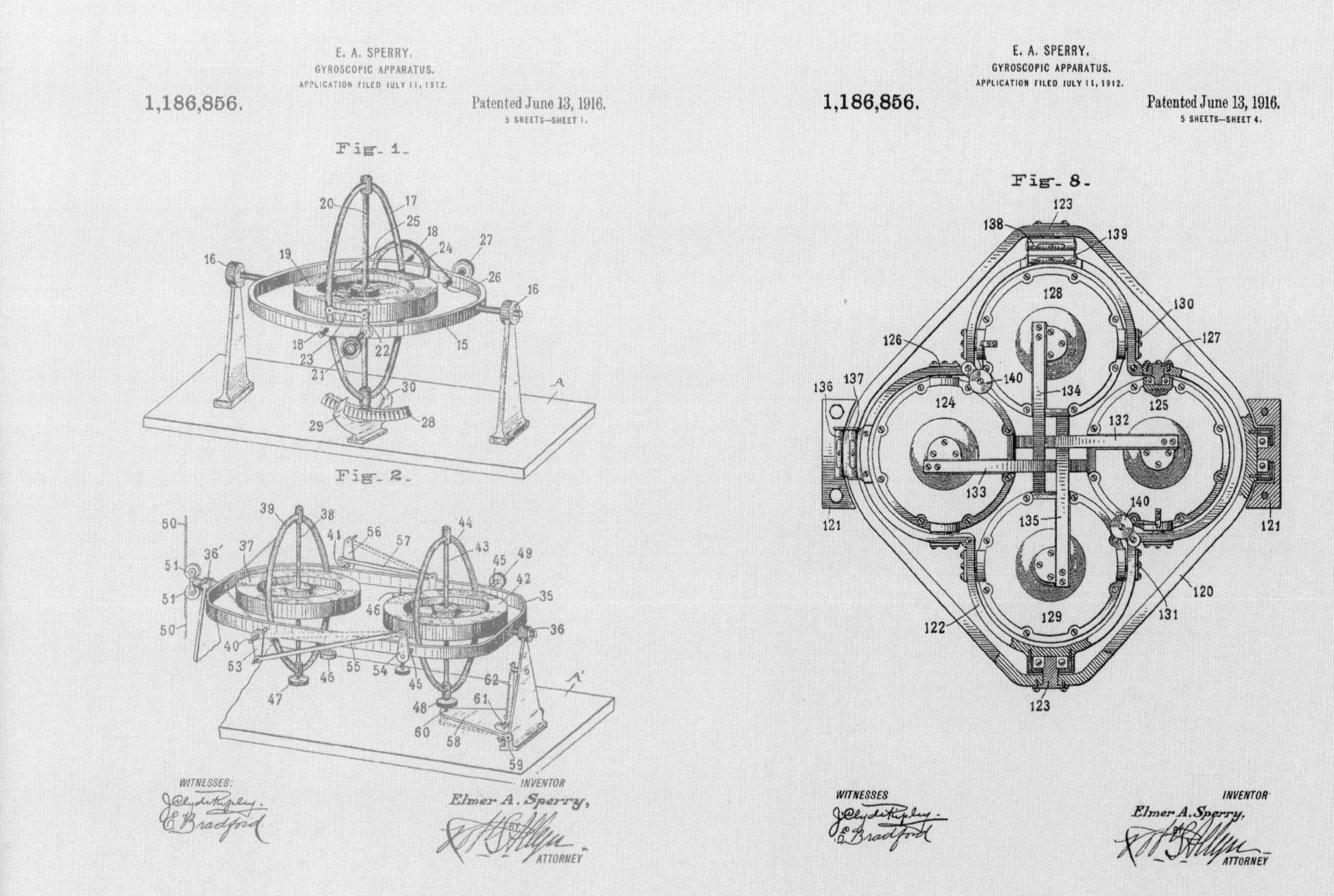

Sperry amassed more than 350 patents before his death in 1930.

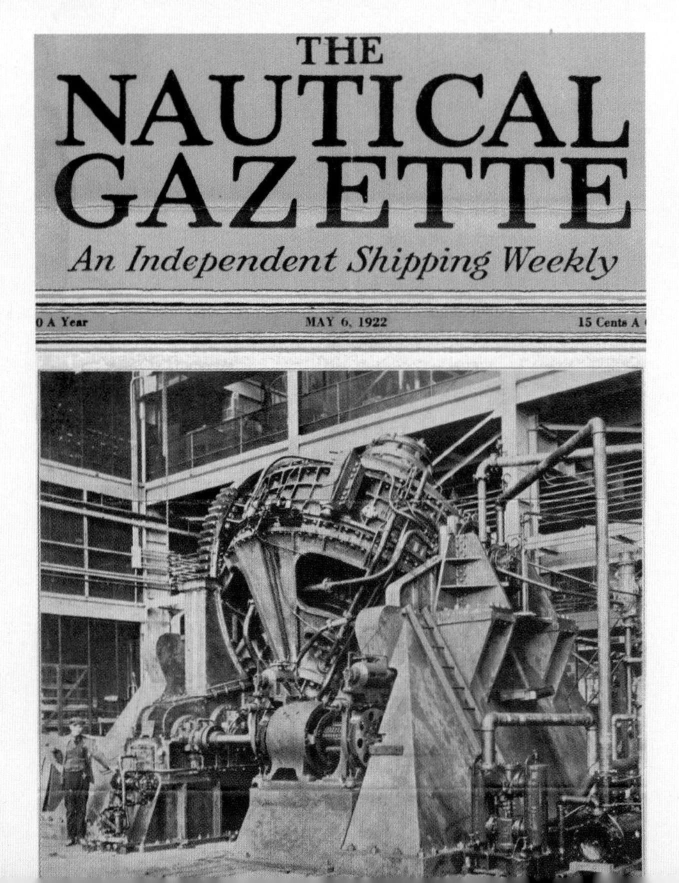
THE NAUTICAL GAZETTE

An Independent Shipping Weekly

0 A Year | MAY 6, 1922 | 15 Cents A

When this gyrostabilizer was built in 1922, it was the largest in the world, with a gyro wheel 12 feet across and weighing 109 tons.

mounted in the center of a ship. When a wave pushed the ship sideways, the gyroscope's precession would lessen the roll.

At least one other engineer had built a similar system, but Sperry added something new: a pendulum that acted as a sensor, picking up the first hint of a roll and activating motors that quickened the gyroscope's response. In 1911, after he had built several models of the gyrostabilizer, the U.S. Navy offered a ship to test a full-size system in. A four-thousand-pound gyroscope was installed, and the refitted ship was launched in 1912. The stabilizer worked well, eliminating minor roll and reducing a 30-degree roll to about six degrees.

At the same time he was designing the gyrostabilizer, Sperry was also working on a gyroscope-based compass. A German engineer concerned with the navigational difficulties of all-metal submarines had already begun pursuing a gyrocompass. Both he and Sperry began from an observation Foucault had made in 1852: that a spinning gyroscope initially aligned toward the north would always point north, even disregarding the rotation of the earth. Transforming this physical principle into a working compass took some engineering tricks, such as limiting its movement to two axes and automatically correcting for shifts as the gyro's position changed with regard to latitude. Sperry applied for a patent on his gyrocompass in 1911 and installed a working model in the same ship as his stabilizer. The two new navigational aids were further refined in the years leading up to World War I; by the time the United States entered the war, many of its ships were equipped with Sperry's inventions.

The problems of navigating ships paled in comparison to those of airplane flight. "Of all the vehicles on earth," Sperry wrote, "the airplane is . . . obsessed with motions, side pressure, skidding, accelerations, pressures, and strong centrifugal moments . . . all in endless variety and endless combination." Early airplanes, such as the Wright Brothers', depended on the pilot to control the plane through physical effort at every moment. This made long flights impossible, since the pilot would grow too tired to wrestle the controls. Another problem was that it was easy for pilots to grow disoriented, especially in clouds or at night, when they had no visual reference to the ground. Sperry used gyroscopes to address both problems.

He tackled the problem of control by designing a four-gyroscope stabilizer. As in a ship, when a plane tipped or rolled, the gyros would exert an opposing force. The plane stabilizer couldn't depend on the large gyroscopes suitable for ships; instead, the movement of small gyros sent signals to motors attached to the elevators and other parts that controlled flight. Sperry's system flew for the first time in 1914; one observer noted that "this marvelous machine can recover lost equilibrium unaided by man." The next year, Sperry began working on gyroscopic instruments that would tell pilots if they were turning or climbing or diving. The gyros in the stabilizer could also indicate the true horizontal level—an "artificial horizon" that no pilot could confuse.

Sperry continued inventing long after he perfected his navigation systems, amassing more than 350 patents before his death in 1930. And while his work had advanced more industries than perhaps anyone except Edison, it was his gyroscopic work that would have the greatest impact. His autopilots, stabilizers, and gyrocompasses are still the basis for marine navigation, and it was the airplane gyrostabilizer, artificial horizon, and other flying instruments that allowed aviation to become an important mode of transportation. In the 1960s, his gyroscopic work was extended even further, as an essential part of the guidance systems in the American space program.

One of Sperry's patents was for this "calculator," a circular slide rule designed to look like a pocket watch.

ENERGY AND ENVIRONMENT

INTRODUCTION

Much of the innovation for which America is famed has its roots in the reductionist approach to problem solving, which involves the inventive mind in identifying a specific issue, or need for improvement, and then focusing intensely on the search for a solution. This kind of thinking has to an extent generated the myth of the lonely genius (such as Leo Baekeland or Steve Wozniak) locked in a garage, surrounded by arcane gadgets, like some medieval alchemist, cut off in his magic isolation from the real world outside.

The first technological apotheosis of reductionism was probably the Industrial Revolution, and its inventors epitomized the isolated noodler at work. James Watt was totally immersed in finding how to keep his cylinder hot while it was being sprayed with cold water to condense the steam inside and cause the partial vacuum needed for piston movement. James Nielson (of hot blast fame) spent years working out the exact temperature of the hot air he wanted to inject into a furnace to obtain maximum smelting performance from inferior coal. Robert Fulton drove himself to exhaustion developing an engine that would coax out the extra knot of speed needed to make his steamboat commercially viable.

On a larger scale, the products of the Industrial Revolution were good examples of both the way in which reductionism isolates science and technology in separate silos, and also of the unexpected and sometimes unwelcome effects that happened when the work of those isolated innovators comes onto the market and interacts with inventions emerging from other, similarly isolated noodlers. The positive side of this interaction—as, for instance, when steam power joined textile machinery to produce clothing everybody could afford; or when mass production generated the first steady year-round jobs and regular wages—was to raise the standard of living of the general populace. The negative side was that industrial innovations such as nineteenth-century production lines and automation brought millions from the villages and farms into the factory towns, and packed them into slum dwellings where they lived ankle-deep in sewage, easy prey to the cholera epidemics that ravaged so much of Europe and the American East Coast in the late nineteenth century. These events raised the consciousness of politicians and innovators and triggered the first of the great public health reforms and the inventions required to carry them out: new medical advances (America opened the first public health lab employing techniques and apparatus developed for the entirely new field of bacteriology), and improvements in sewerage technology (in Victorian England: Doulton's first glazed sewer pipes connected to Jennings's first patent ceramic flush toilets).

The modern world faces a continuation of this dual nature of innovative social effects, but on a global scale. Thanks to the twentieth-century drive to spread industrialization across the globe and then to raise world standards in public health, the planet has come under sustained pressure from both the effect of pollution and of uncontrolled population growth. A recent guess estimated one effect of this phenomenon to be a loss of 6,000 species every day, and forests are disappearing at the rate of one Switzerland-equivalent per year. Population levels are forecast to hit unsustainability before mid-century. Add ozone layer depletion and possible global warming to the mix, and the full-scale ramifications of what can sometimes go wrong with the reductionist approach become apparent: We may have spent several centuries not seeing the wood for the trees. Although reductionist innovation has brought most of us the highest material standard of living in history, it has done so only at a considerable, perhaps unacceptable environmental price.

One of the fundamental effects of invention has been to introduce once-for-all change. Innovation can never be wished out of existence. We live with the tools for change that our inventors gave us. In the long run, this has meant that some of the greatest inventors in history caused problems they could never have foreseen. But in the main, they also gave us the tools with which to find solutions to those problems.

This group of innovative thinkers represents a growing number of people concerned to apply this principle and to encourage us to "think globally, act locally." Their efforts range from pollution-free forms of energy, to environmentally friendly textile manufacture, to cheap water purification. Given the scale of our present worldwide ecological difficulties, there is still much to do.

—JB

GEORGE WASHINGTON CARVER

George Washington Carver was many things to many people. As someone who transformed himself from an orphaned ex-slave into a scientist and national hero, he was a living example of the American Dream. As a devout Christian, he was a symbol of religious faith. As one of the most famous African Americans of his time, he represented the possibility of advancement for people of any color. Carver's most important role, though, may have been as an agricultural scientist. Working at the Tuskegee Institute in Alabama, he helped revolutionize farming in the South, introducing alternative crops, the idea of crop rotation, and natural methods of fertilization, as well as inventing hundreds of products and uses for his favorite plants, the peanut and the sweet potato.

Carver, born in 1864, had an early childhood full of conflict and tragedy. His mother, Mary, was a slave owned by Moses and Susan Carver, farmers living near the small Missouri town of Diamond. Carver never knew his father, who is said to have died in an accident before his child's birth. Although the Civil War ended in April 1865, many rural areas of the South remained in turmoil. Sometime that year, a band of marauding "bushwhackers" kidnapped Carver and his mother and carried them off to Arkansas. Moses Carver sent an emissary to ransom them, although by the time he arrived Mary apparently had been sold and had disappeared; the young Carver was traded for a $300 racehorse.

After George was returned, the childless Carvers adopted him and his brother. Because he was sickly, George did chores around the house rather than work in the field. He also spent a great deal of time outside, where he developed his love for plants. In 1897, he remembered this time of his life: "Day after day I spent in the woods alone in order to collect my floral beauties and put them in my little garden I had hidden in brush not far from the house." Carver's skill at caring for flora soon earned him the nickname "plant doctor," and people would bring him their sick plants for treatment.

Carver's early education came from his adoptive mother and a single book, *Webster's Elementary Spelling Book*, which he read so many times, he later said, "I almost knew the book by heart." The local school would not ac-

When George Washington Carver joined Tuskegee University in 1896, most Southern farms grew only cotton. Carver championed alternative crops such as peanuts and soybeans, whose uses he demonstrated in his lab.

cept black students, so when he was 12 Carver moved to the nearby town of Neosho, Kansas, to begin school. There he lived with a black couple, Andrew and Mariah Watkins, and earned his keep by doing domestic chores and helping Mariah with her laundry business. After a year in Neosho, he went on to attend schools in several Kansas towns, eventually finishing high school and moving to Kansas City to learn shorthand and typing at a business school.

Clerical work could not satisfy Carver's thirst for knowledge, and in 1885 he applied to a college in Highland, Kansas. He was accepted, but when he showed up for class, school officials were shocked to find out that he was black, and they refused to let him in. A few years later, he moved to Winterset, Iowa, where he set up his own laundry business and became close friends with a white couple who urged Carver to try college again. In 1890 he entered Simpson College, where he discovered his talent for painting. A year later, his art teacher, who had noticed his detailed drawings of plants, persuaded him to study agriculture at what is now Iowa State University. He was the school's first African American student.

With his natural talent for plants, Carver thrived at Iowa State. After he graduated, he stayed on to teach and to study for his master's degree. He might have remained there if not for Booker T. Washington, a leader and educator who had founded the Tuskegee Institute, the leading college for African Americans. When Washington invited him to join the faculty at Tuskegee, Carver saw it as an opportunity to serve his race. "It has always been the one great ideal of my life to be of the greatest good to the greatest number of 'my people,'" he wrote Washington. "This line of education [agriculture] is the key to unlock the golden door of freedom to our people."

At the turn of the twentieth century, Southern agriculture was devoted primarily to cotton. But cotton is a destructive crop, draining nutrients from the soil. Also, any region devoted to one crop is at a greater danger of agricultural disaster, from plant diseases or insects. After more than a century of cotton growing, much Southern farmland was growing less productive, and many fields were ruined. This affected African American farmers more severely, since

Encouraged by an art teacher who had noticed his drawings of plants, Carver began studying agriculture at Iowa State College in 1891. He was the college's first African American student.

While pursuing his master's degree at Iowa State, Carver was approached by Booker T. Washington, the founder of the Tuskegee Institute, the country's leading school for African Americans. In 1896, Carver accepted a position on Tuskegee's faculty, pictured here. He remained at Tuskegee for four decades. Carver is at the far left of the top row.

their farms were almost always smaller than those of whites, and were often rented rather than owned.

Carver thought that a scientific knowledge of agriculture could help meet these challenges. He also had a deep belief that man should work in concert with nature, paying attention to natural processes and being careful not to waste resources. At Tuskegee, Carver started by experimenting with new ways to work the local land, which was in poor condition, then began using the school as an educational pulpit to encourage local farmers to follow his example.

During Carver's years at Tuskegee, he promoted three agricultural innovations: organic fertilizing methods, crop rotation, and new crops that returned nutrients to the soil. By plowing the unused remains of a harvest under the soil, a farmer could create a kind of natural compost heap that acted as a highly effective fertilizer for the next crop. Rotating crops—alternating cotton and other crops from year to year—reduced cotton's destructive power. Carver's real passion, however, was exploring alternative crops: soybeans, pecans, the sweet potato, and, above all, the peanut.

The worst thing that cotton did to soil was rob it of nitrogen, an essential nutrient. Carver knew that peanuts and sweet potatoes replenished nitrogen; rotating these crops with cotton could heal the South's fragile soil. It would also lower the danger that insects like the boll weevil could destroy the region's economy.

Many farmers began testing Carver's ideas in their fields and found that he was right: Peanuts and sweet potatoes did restore the soil. But while the market for cotton was well established, demand was limited for these new crops. So in his laboratory Carver set to finding practical uses for them. Over the course of his four decades at Tuskegee, he created hundreds of products based on peanuts and sweet potatoes. From the peanut, he was able to make plastics, animal feed, oil, ink, paper, and stains, as well as many foods, from cheese and milk to meat substitutes, candy, and breakfast snacks. And from sweet potatoes he devised methods for making flour, glue, alcohol, syrup, and starch, among other things. In 1921, he was called to testify before the U.S. Congress when it was considering imposing taxes on peanut imports. He dazzled the legislators by exhibiting some of his peanut products, including mock oysters and a face cream.

Carver patented only three of his agricultural products: cosmetics, paints, and stains made from soybeans. His goal in agricultural research was not to get rich—he lived modestly and made a point of refusing money offered to him—but to improve the plight of farmers, especially African American farmers. A deeply religious man, he credited God for his many inventions, and would say, "God gave them to me. How can I sell them to someone else?"

His appearance in Congress and many favorable articles about him made Carver a famous man. In the 1920s, he began speaking to white youth groups in the South to improve race relations. As his amazing life story became widely known, he became an American folk hero.

When Carver died in 1943, his entire estate was given to the Tuskegee Institute to start the Carver Research Foundation for agriculture. Another permanent reminder of his life was established that same year. Just six months after he passed away, President Franklin Roosevelt dedicated the George Washington Carver National Monument, near Carver's childhood home in Diamond, Missouri. The honor was unprecedented—not only was it the first national monument to an African American, it was the first to anyone other than a U.S. president.

Among the hundreds of products Carver coaxed from peanuts and soybeans were dyes, paints, cleansers, charcoal, and fuel, as well as many different foodstuffs.

CARL DJERASSI

The world does not lack brilliant organic chemists, but few have had such a profound effect as Carl Djerassi. His synthesis in 1951 of artificial progesterone earned him the title of Father of the Pill. Before that, Djerassi had been instrumental in the creation of the first antihistamine and the first synthetic steroids. And his innovation continued when he became a director of research and a chemical entrepreneur, leading a team of scientists exploring the hormonal secrets of insects. This work led to new approaches to pest control that relied on mimicking natural chemicals rather than on dangerous neurotoxins.

Djerassi's interests outside the lab are even more diverse. He is a well-known collector of art, especially that of Paul Klee, as well as a patron who runs an artists' and writers' retreat on his ranch outside Santa Cruz, California. In the past few years, Djerassi has been breaking new educational ground by teaching scientific ethics through fiction writing, specifically the writing of "science in fiction," stories and novels revolving around the processes and customs of the scientific establishment. His own science-in-fiction novels have been well reviewed and well read; the first, *Cantor's Dilemma*, a novel about the fierce competition that drives star scientists, has sold more than 100,000 copies.

Djerassi, who was born in Vienna in 1923, had an eventful adolescence. His father was a doctor and his mother a dentist, although she, too, had been trained as a physician. Djerassi spent his first few years in Bulgaria, where his father was from, then returned to Vienna for his education. (The move back to Vienna was also due to his parents' divorce.) He was doing well in school and planning to go into medicine like his parents when, in March 1938, Nazi Germany occupied Austria. While many Austrian Jews thought they would be safe in their native country, Djerassi's father knew to take quick action. He temporarily remarried his ex-wife, to get her and his son Bulgarian passports. As foreigners, they could leave Austria; his mother traveled to England, while Djerassi went to Bulgaria with his father, where he enrolled in the American College (a high school) there. In September 1939, just after Germany invaded Poland, Djerassi and his mother were granted visas to the United States, and

Djerassi at his Syntex lab in Mexico City in 1951, working with his assistant Arelina Gonzalez on synthetic progesterone.

Carl XVI Gustaf, King of Sweden, at center, visited Djerassi's insect-control company Zoecon in 1984. Djerassi looks on as Gerardus Staal, Zoecon's director of biological research, shows the king two immobilized cockroaches.

In 1951, Carl Djerassi (center, seated) was the first person to make synthetic progesterone—a hormone that eventually led to the birth control pill. At a press conference, Djerassi and his colleagues examined the yamlike plant from which they extracted the chemical precursor to the hormone.

he pulled up roots again; the two of them soon arrived in New York, virtually penniless.

In America, Djerassi and his mother were again separated—she went to work in upstate New York, while he was taken in by a generous family in Newark, New Jersey. There he began his American education at a junior college, although he had not completed high school. Djerassi soon realized that he would need a more substantial education to get anywhere. He had no financial resources, so he took what he thought was a logical step: He wrote to Eleanor Roosevelt asking for help. Improbably, it worked. His letter was forwarded to the Institute for International Education, which secured him a scholarship at a tiny Presbyterian college in northwestern Missouri.

For a while, Djerassi supported himself by speaking at local churches on the conflict in Europe. The next year he moved on to Kenyon College in Ohio, where the 17-year-old set out to get his bachelor's degree in chemistry in just one year. Carrying heavy course loads through two semesters and a summer, Djerassi earned his degree in this seemingly impossible time frame.

The new graduate applied to many of the chemical companies in New Jersey (the center of the industry in the United States), and was soon hired by the Swiss pharmaceutical firm CIBA. Djerassi's rise was swift; during his first year there, he was involved in the creation of the first antihistamine and was credited as a coinventor on the patent and a coauthor of the article detailing the work. CIBA was also a leader in steroid research, and exposure to that science in the lab, as well as his reading of a seminal book on the subject, soon had Djerassi hooked. A disabling knee injury kept him out of World War II, allowing him to pursue a Ph.D. at the University of Wisconsin, which he earned in just two years.

Djerassi returned to CIBA for a few years, but in 1949 he was offered the chance to head a steroid research group in the Mexico City office of the Syntex company. At the time, Syntex was involved in the race to make a synthetic version of cortisone. Within two years, Djerassi and his colleagues had done it first, putting his Mexico City lab on the steroid research map.

The next challenge Djerassi turned to was making a form of progesterone that could be taken orally. Progesterone is produced naturally when a woman is pregnant; it prevents her from ovulating. In the 1920s, an Austrian endocrinologist had proposed that this hormone could be used as a contraceptive. But that is not why Djerassi was trying to synthesize it. Syntex was already the largest producer of progesterone, which was used to treat menstrual disorders and was thought to be a treatment for cervical cancer. But the hormone could be administered only by injection; this made it an expensive and not always practical treatment and limited its uses. An orally delivered version would avoid those problems.

On October 15, 1951, Djerassi's lab converted a precursor to synthetic progesterone—extracted from a yamlike plant—into norethindrone, which turned out to be much more active than the natural hormone as well as capable of being taken orally. "That was the Eureka part," Djerassi says. The lab's published research made its way to Gregory Pincus, a biologist looking into contraception, and John Rock, a doctor at Harvard. Using Djerassi's synthetic hormone, they developed the oral contraceptive pill, usually known as the Pill. Since its introduction in 1960, according to one of its manufacturers, some 80 percent of American women have used it.

By the mid-1960s, Djerassi's lab began investigating insect hormones. It was an exciting time for this research: Improved chemical and analytical techniques were allowing scientists to deeply explore insect endocrinology for the first time. One of the most exciting discoveries of the time was that a single hormone, ecdysone, governed insects' molting process. Syntex and several other labs worked to synthesize this complex steroid; Djerassi's team was one of the first to do it. (By this time Djerassi was the director of research at Syntex; he monitored others' work and assigned scientists' priorities, but he didn't do the actual synthesis.)

That success is "really what got us interested in insects," Djerassi explains. "But it was a chemical interest rather than biological." Eventually, he continues, having done this work, "we wondered, can we use it for something?" Djerassi hired two of the most distinguished endocrinologists in the United States, Herbert Röller and Carroll Williams, and began exploring what applications the research might have.

A discovery by Röller showed the way. He had found a "juvenile hormone" in insects that maintained their larval state. When this hormone is "turned off," the insect grows into an adult and is able to reproduce. If one could synthesize this hormone and continuously apply it, insect pests would never live to reproduce. "It would be the equivalent to using a human contraceptive in which you prevent the onset of puberty," Djerassi has noted. The lab was able to synthesize this hormone, and in 1968, Djerassi convinced Syntex to spin off this research division as a separate company, which he named Zoecon.

At the time, the pesticide DDT was being banned worldwide, even though, Djerassi notes, it had done much good for humans, if not for animals. "Chemophobia is to a very large extent propagated by the word pesticide, in its dirtiest sense," he says. Zoecon was formed to find alternative, "biorational" approaches to pest control, methods that would kill the insect but not humans or other, nontargeted animals. "Conceptually," Djerassi explains, "our approach to insects was very similar to Syntex's earlier approach to birth control."

By the early 1970s, Zoecon had developed a product, Altosid, based on insect juvenile hormones, which are part of a larger class of agents called insect growth inhibitors. "In many respects it was one of the ideal pest-control agents," Djerassi says, "in terms of being biodegradable, and being metabolized completely through innocuous things." Altosid was adopted in the fight against insects that do their damage as adults—fleas, cockroaches, and, most important, mosquitoes. Today, Altosid is common in many mosquito-control products. But Djerassi laments that most money for insect-control research, development, and treatment goes to commercial crop pests. "Malaria is one of the greatest killers of human beings," he says, "but there really is not much money in public health things."

For the past decade, Djerassi has dedicated much of his time to writing novels and stories that explore the culture and personalities of science as it is actually practiced. He has also begun teaching a writing seminar at the Stanford University School of Medicine, where prospective doctors can explore through fiction the ethical dilemmas of their future practice that often, according to Djerassi, "are not raised for reasons of discretion, embarrassment, or fear of retribution." Djerassi's latest novel, *No*, was recently published in paperback, and he is preparing a new book, *One Man's Pill*, an exploration of how his contraceptive discovery changed both him and the world.

SALLY FOX

Sally Fox didn't invent colored cotton—it has always existed in nature. The Incas used it in their textiles. Khaki was first made in India from brown cotton. In the United States, slaves grew brown and green cotton in their own gardens, since they were forbidden from growing white cotton that they might sell.

When Fox introduced her own colored cotton to the world in 1989, though, she had done something no one had thought possible: created a naturally colored cotton that could be spun (made into thread) on a machine. Until then, colored cotton had suffered from short, weak fibers, which meant it had to be spun by hand, a slow and expensive task. The only cotton that was commercially viable was white cotton, which had been bred and refined to have long, strong fibers, perfect for machines to work with. But white cotton is not an environmentally friendly product. Before it becomes jeans, or sheets, or a shirt, it has to be bleached and then dyed, and both processes create large amounts of pollution. Fox Fibre, the brand name of Fox's cottons, required no bleaching or dyeing. Her first two colors—a celadon green and a warm reddish brown—were immediate sensations. For the first time, the textile industry and the public saw how environmentally friendly clothes could be. Orders for her cotton began pouring in.

Fox never imagined she would be a cotton revolutionary. But in many ways everything she had done before announcing her new cotton had been perfect training for the role. When she was 12, she fell in love with the process of spinning. With a spindle she bought with baby-sitting money, she created thread out of all sorts of materials, from the cotton in medicine bottles to her dog's hair. She was soon a master spinner. But Fox also dreamed of living and working far from cities. Until she was seven, she lived on the last acres of land her father's family had homesteaded in Woodside, California. "There were big expanses of land," she says. "When we moved from that into Menlo Park I never got over it. I always wanted to live in the country and have some means of supporting myself."

In high school, she discovered a new passion: entomology, the study of insects. An entomologist visiting from Kenya, Elizabeth Wangari, taught a

Each year, Fox examines the her main crops and test fields to determine which plants exhibit new qualities and which should be selected for the next year's planting.

By carefully selecting the best plants of a cotton crop, Sally Fox gradually bred a new kind of cotton: colored and spinnable.

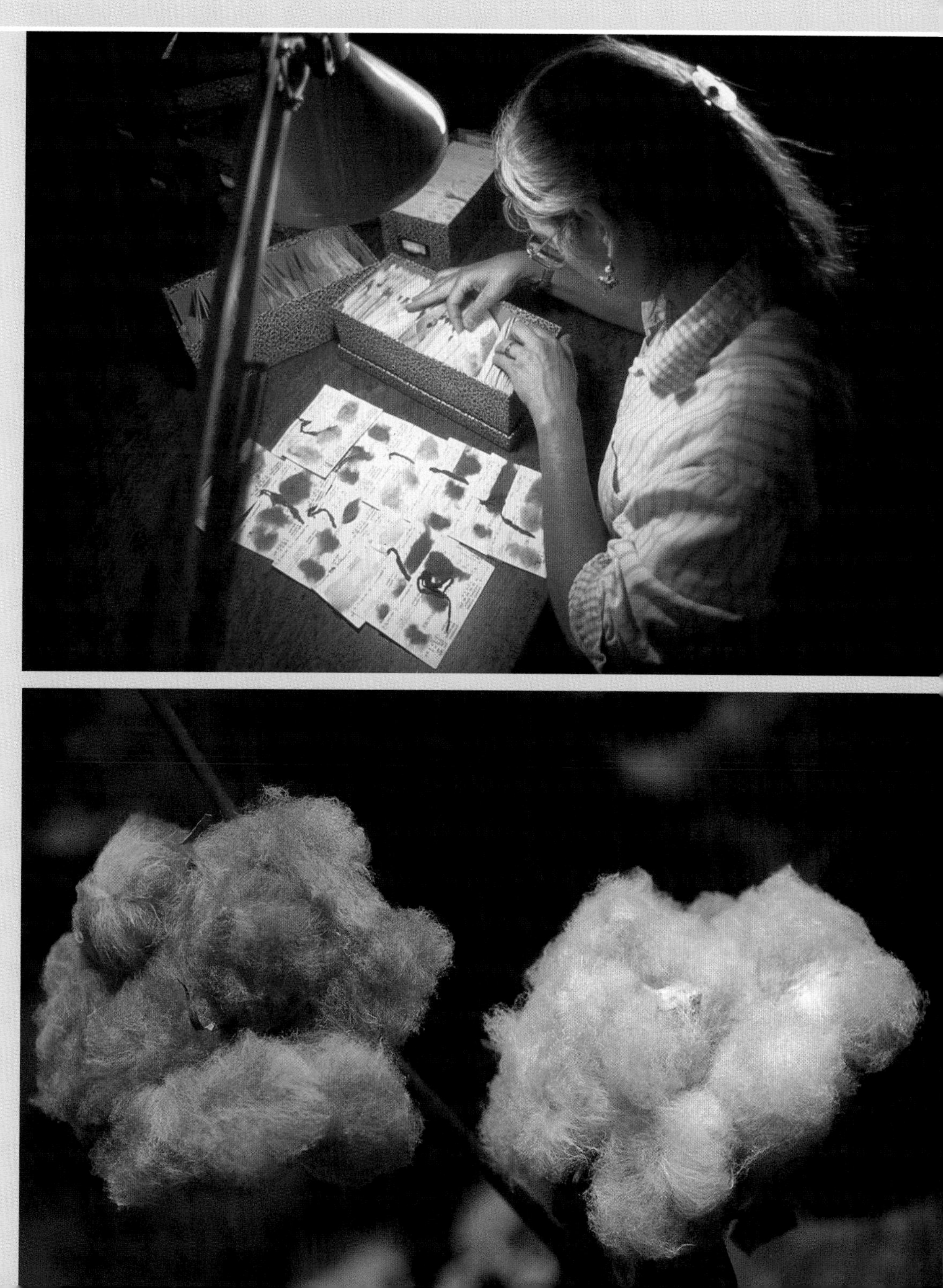

ARTIFICIAL SELECTION AND THE ORIGIN OF FOX FIBRE

The process of artificial selection in plants—choosing, over many generations, the most beneficial or fertile or beautiful examples—is as old as agriculture. Long before the age of industrial farming, all crop plants were the product of countless generations of careful selection, with farmers choosing to plant seeds from the strongest or most bountiful of their plants. Naturally occurring corn, for example, has an ear only an inch or two long, with small kernels; it was only through selection and cross-breeding over centuries that our modern food staple was created.

Although artificial selection has been around for millennia, the science behind it—genetics—was discovered only in the nineteenth century, by the Austrian botanist and monk Gregor Mendel. In carefully studying garden peas in his nursery, Mendel noticed that certain differences between generations—for example, in plant size or seed shape—were predictable, and he proposed that these traits were controlled during reproduction. Some differences had only two states—shrunken or nonshrunken, for instance—while others had more subtle variations. It is understood now that simple differences are controlled by one gene, while subtler changes are the result of combinations of genes. By breeding pea plants with specific characteristics—and isolating them from other plants—Mendel could produce plants that emphasized those qualities.

In her pursuit of machine-spinnable, naturally colored cotton, Sally Fox used the same basic ideas that Mendel developed (and which have been refined in the ensuing 135 years). Fox estab-

class on the subject and helped Fox get an internship at Zoecon, Carl Djerassi's company, which was developing natural ways to control insects. Wangari also encouraged her to go to college—Fox had wanted to start a hand-spinning business—where she studied biology and entomology.

After she graduated, Fox joined the Peace Corps and traveled to The Gambia, in West Africa, to help fight the pests and diseases that affected rice and peanuts. What she really learned there, though, was the dangers of pesticides. Europe and the United States had recently banned DDT and other chlorinated-hydrocarbon-based pesticides, and some European manufacturers had "donated" their large stocks to African countries—in unlabeled, leaking 55-gallon drums. It was an environmental disaster waiting to happen. The local men hired to administer the pesticide were instead selling bags of it at the market for people to use on household pests. Fox began giving safety classes on the pesticides' use, but she also became very ill from exposure to them. "It was so horrible that I actually had to leave" before her two-year assignment was finished, she says. "So I'm sort of a fanatic against pesticides."

When Fox began looking for work in the United States in the early 1980s, the farming industry had just entered a long economic depression, from which it has never fully recovered. The only job Fox could find in her field was as a pollinator for a cotton breeder searching for pest-resistant plants. She was bored by the job, since it took a year to see if new strains of cotton were better than old ones. One day she found a bag of seeds that produced cotton that was pest-resistant but brown. The breeder hadn't pursued these seeds, because he didn't think he could get rid of the color. Fox was intrigued. "I said, 'Why aren't we doing this?'" she remembers. "And he said, 'Why don't you do it?' So I went through the seeds and hand-spun each single one. I decided which ones were the easiest ones to spin and I planted those. That was the beginning."

For the next seven years—even after she stopped working for the cotton breeder—Fox tended her cotton plants, now growing in pots on her back porch. Each year, when the cotton bolls opened up, she would carefully select seeds from the plants with the best fibers and the best colors. She also crossbred her cotton with white cotton, to produce a longer fiber, or staple. "I was an entomologist, and I hadn't read all the plant books saying you couldn't breed for longer staple," she says. Eventually she had two colored cottons that were stable—they didn't change when planted in the field—and were spinnable.

lished several characteristics she was looking for, including color, the length of the cotton fiber, the softness of the fiber, and the strength and pest-resistance of the plant.

Over a period of 10 years—or 10 generations of plants—she examined the cotton for these characteristics. Each generation's best plants were then bred, through pollination, with each other. The following year, a few of the plants would exhibit stronger qualities; those, in turn, would be selected out, bred, and grown. At a genetic level, each generation of these hand-selected plants had a combination of genes that better produced the characteristics Fox was looking for. Some of these qualities were linked—pest resistance always came with certain colors of cotton—while others were the result of combinations of independent genes.

Eventually, the qualities of Fox's plants were sufficiently different from their ancestor that she could essentially patent them and call them her own.

Genetic engineers are also working toward many of these same goals, such as stronger plants and bigger fruits—but they are using a much faster, although less well explored, means. By incorporating the actual sequences of DNA that code for beneficial characteristics in a certain plant—strawberries' frost resistance, for example—into another crop, they can speed up the processes of artificial selection and cross-breeding a thousand-fold and, in many cases, produce plants that no amount of breeding could ever make. These techniques, pioneered in agriculture, are also being applied to animals and even humans, in the budding science of gene therapy.

She applied for and won Plant Variety Protection Certificates for them, the plant equivalent of patents.

In 1989 she sold her first crop, 122 bushels of cotton, to a Japanese mill. American customers weren't far behind. Levi's began making "natural" jeans and other clothes. L.L. Bean, Land's End, and Esprit also placed big orders. Fox was running a $10 million business. She put together a network of cotton growers that produced hundreds of thousands of pounds of Fox Fibre cotton.

But Fox's natural cotton revolution has not been a smooth one. California's powerful cotton growers were afraid that her colored varieties would contaminate their own crops. They imposed strict rules on her operation, which forced her to move to Arizona in 1993. Six years later, Arizona cotton growers did the same thing, and Fox had to relocate again, this time to Northern California. Even worse for her business were the changes in the spinning industry. Between 1990 and 1995, most of the spinning mills in Japan and Europe closed, as well as many in the United States. "What was going on was the beginnings of globalization," Fox explains. "Everything was moving to Southeast Asia and South America." These mills wouldn't or couldn't process the relatively small quantities of cotton her farmers produced, and she lost her big customers.

Until the spinning industry moved to less-developed countries, Fox's cotton had a financial advantage over traditional cotton. It cost about $2 per pound of cotton to treat and dispose of the toxic waste from the cotton-dyeing process. However, many of the new, offshore mills weren't required to protect the environment, and didn't. On a recent trip to Hong Kong, Fox noticed a headline in the local newspaper: "Farmers Downstream from Denim Plant Lose Entire Crop." A dye plant had dumped its waste just when the rice harvest began, and killed some 300 acres of crops. "If it kills the plants like that," she asks, "What are the long-term effects?"

Through all of her cotton's ups and downs, Fox has continued to develop new colors. It takes about 10 years of careful selection from generation to generation to coax a usable variety of cotton from an initial cross-bred seed. "You get a color, and it's not a great plant, and it's not a great fiber, but it's a color," she says. "And then you keep at it until you get a better plant. You just work at it year after year." She has a new redwood color almost ready for market; combined with white cotton, it makes a shade of pink. She's also added a deeper "New Green" and the chocolate-toned "Buffalo" to the Fox Fibre line.

Fox now concentrates on smaller mills and smaller customers, and she is rebuilding her business and her network of growers. One new client is emblematic of her hopes for Fox Fibre. Two brothers in Mexico, she explains, want to rebuild their family's spinning business. Their grandfather had built a mill, then a dyeing plant, then a factory, and finally a store for the clothes he produced. By the time the brothers were in their twenties, though, the village's water supply had been destroyed by the dyes, and the mill shut down. With Fox's cotton, they hope to start the mill up again and to set an example for other mills in the area that it's possible to make textiles without destroying the local environment. "They're the kinds of people I dream of selling to," Fox says. "It can really make a difference."

Selective breeding, a process that is thousands of years old, requires careful attention to each plant and a knowledge, acquired through experiment, of how plants can be cross-bred.

R. BUCKMINSTER FULLER

In 1927, R. Buckminster ("Bucky") Fuller was a failed businessman who was drinking too much, depressed about the death of his first daughter, and worried about how to take care of his family. That winter, while living in Chicago, he walked out to Lake Michigan to throw himself into its icy waters. "I said to myself, 'I've done the best I know how and it hasn't worked. I guess I'm just no good,'" he remembered. Fuller gave himself a choice: jump or think. He stood on the lakeshore for hours, finally deciding that he didn't have the right to kill himself. And so began Fuller's career as an inventor, thinker, and futurist.

From that point on, Fuller thought of his life as an experiment designed to discover what he could "do effectively on behalf of all humanity that could not be accomplished by great nations, great religions or private enterprise." He christened himself "Guinea Pig B" (B for Bucky) and started "thinking about our total planet Earth and thinking realistically about how to operate it on an enduringly sustainable basis as the magnificent human-passengered spaceship that it is." The results of Fuller's lifelong experiment include the geodesic dome, new types of houses and cars, and a new kind of geometry. Underlying all of his work was the profound realization he had come to that night in Chicago, that he could "find ways of giving human beings more energy-effective" systems and machines that would create a higher quality of living for everyone. "Under those more favorable physical circumstances," he wrote, "humans would dare to be less selfish and more genuinely thoughtful toward one another."

Anyone who had watched Fuller growing up would have been surprised at his desperate circumstances in 1927. Richard Buckminster Fuller, Jr., was born in 1895 to a wealthy and long-established Massachusetts family. Although his father died in 1907, Bucky had an active and fairly happy childhood.

Fig:5.

INVENTOR
RICHARD BUCKMINSTER FULLER
BY

The family spent its summers on a private island off the coast of Maine, and as a teenager he attended Milton Academy, a prestigious prep school. In 1913, like four generations of Fullers before him, he entered Harvard University.

Fuller never really fit into Harvard's rigid social structure, and he earned only a C average. At the end of his freshman year, he took the college money his mother had put away and went to New York City, where he spent all of it and more trying to romance a Ziegfeld Follies dancer. The university expelled him for "irresponsible conduct."

After getting kicked out of Harvard a second time, Fuller met and married Anne Hewlett, who also came from a venerable family. During World War I, his mother secured him an officer's commission in the navy by donating one of the family's boats. Fuller loved the navy; he patrolled the northeastern coast for much of the war, and later designed a mast-and-winch system to rescue seaplane pilots. After the war, he began working with his father-in-law on a concrete-block building system. Although their investors had high hopes, the American construction industry stuck to traditional techniques, and in 1927, with the company failing, Fuller was forced to quit. He was a new father, jobless, and broke when he walked out to Lake Michigan.

After his epiphany, Bucky Fuller resolved to think before he did anything. For a year he kept silent, speaking only when he was sure he had something important to say. He thought about what he could do to help people live better. He started with something he already knew a bit about, and which was also a basic human need: building shelter. And he had an idea for a new way to build. Fuller knew that while conventional buildings' strength came from compression—stacking bricks on top of each other—tension was far more efficient. Steel, for example, can support twenty times more weight suspended from it (as with the steel cables of a suspension bridge) than placed on top of it.

In 1929, Fuller unveiled the Dymaxion House, a residence unlike anything that had ever been built. (An advertising writer dreamed up the word Dymaxion, from "dynamic," "maximum," and "tension.") At the center of the house was a tall pole, the "mast," from which the body of the six-sided house

In 1959, Buckminster Fuller and his wife, Anne, built a geodesic dome that served as their home in Carbondale, Illinois, where he taught at Southern Illinois University.

One of the largest geodesic domes in the world was built as the United States Pavilion for Expo '67, a world's fair held in Montreal. Today, the dome houses Biosphere, an ecologically oriented science museum.

was suspended by steel cables. Inside, the space was split up into traditional, if oddly shaped, rooms: bedrooms, kitchen, living room. Fuller also included a "go-ahead-with-life" room, which incorporated a library, a radio, a television, a typewriter, and a drawing board. In his book *Bucky Works*, Fuller's longtime colleague J. Baldwin calls this room "nothing less than a personal multimedia center—in 1927!"

Fuller's Dymaxion House was designed to be a portable residence that a family could buy once (for approximately the cost of a high-end automobile), and then easily pack up and take with them when they moved. It was meant to be built from aluminum, a light, strong, and durable material. The whole thing would weigh only three tons, compared with the 150 tons of a normal building, and its hexagonal shape and the strength of steel tension would keep it sturdy and safe. While Fuller's radical house quickly captured the public's attention, the building industry was less enthusiastic. Because it was easy to build and used a minimum of standardized components, the Dymaxion House would be very affordable, but only if it were mass-produced, and mass production was the enemy of architects and builders. When one (and as it turned out, the only) Dymaxion House was built in 1945, thousands of orders poured in. But Fuller was never an effective businessman, and his disagreements with stockholders and manufacturers doomed the project.

After the failure of the Dymaxion House, Fuller went back to doing research. He was thinking about geometry, specifically about a geometric system he had begun formulating in the 1930s. Just as he had done with his own life in 1927, Fuller was determined to start over from scratch. Instead of exploring the conventional—Euclid's abstract ideas of points and straight lines—Fuller looked to nature, examining how natural shapes can be broken down into smaller and smaller elements. The smallest element he ended up with was the tetrahedron—a pyramid with four equal sides, the most minimal enclosed space possible. "The tetrahedron is the basic structural system," he wrote, "and all structure in the universe is made up of tetrahedronal parts." From this starting point, Fuller constructed a new kind of geometry, a system he called "synergetics" (from "synergy" and "energetic").

In 1948, Fuller was looking for a way to apply these principles to making a building, and began thinking about making a dome-shaped structure. Humans had been using domelike buildings such as igloos and yurts for thousands of years. Fuller thought that a dome made with the tetrahedron as its basic building block would "have structural strength without any

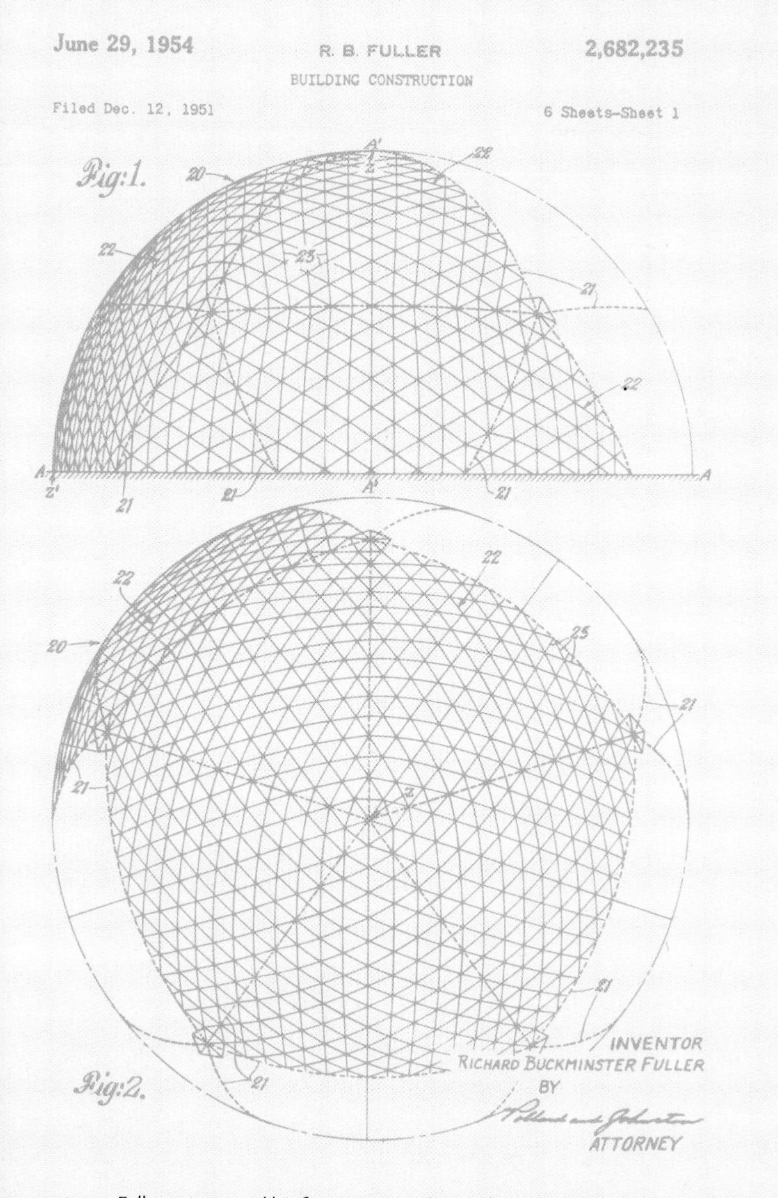

Fuller was granted his first patent on the geodesic dome in 1954. In the application, he detailed the construction of "a frame of generally spherical form in which the main structural elements are interconnected in a geodesic pattern."

Fuller's first large dome was built for the Ford Motor Company's headquarters in Dearborn, Michigan, in 1953. The dome, which protected a round courtyard, weighed eight and a half tons—versus the 160-ton conventional structure Ford's engineers had proposed.

A MOLECULE FOR BUCKY

In 1985, three scientists, Robert Curl, Harold Kroto, and Richard Smalley, stumbled upon a molecule no one had seen before. In Kroto's investigations of how chains of carbon were formed in deep space, he enlisted the help of Smalley and Curl, who were experts in exotic chemistry. While experimenting with using a laser to vaporize carbon, they were left with some intriguing soot. At the conclusion of the experiment, they found that they had created a stable cluster of 60 carbon atoms. What was most intriguing about this cluster was its shape: The carbon atoms formed a spherical cage made of up hexagons and polygons. The sphere had a close resemblance to both Buckminster Fuller's geodesic domes and a soccer ball. The new molecule was officially named buckminsterfullerene, but it has a nickname that reflects both of its real world analogs: Buckyballs.

The discovery of fullerenes was exciting for several reasons. They represented a completely new form of carbon, in addition to diamond and graphite. A new form of molecule, especially one as prominent as carbon, promised all kinds of new and unforeseen applications. And Buckyballs were beautiful—symmetrical and strong, they seemed tantalizing proof of some kind of harmony in the design of the universe.

When other researchers found an easier way to make fullerenes in 1990, the race was on to find applications for the new molecule. Because fullerenes are round, some scientists thought they could serve as a lubricant, with each molecule acting as a ball bearing. Others looked into using Buckyballs as ion-rocket fuel and as superconductors.

A decade later, there are still no Buckyball-laden products on our shelves. But scientists have continued to open up new potential fields for the molecule. Current research is exploring their use in therapies for HIV (a modified buckyball fits perfectly into an HIV enzyme) and ALS, Lou Gehrig's Disease (buckyballs act as powerful antioxidants, slowing the progress of the fatal disease).

Research into fullerenes also led to the discovery of another form of carbon molecules: nanotubes, long chains of carbon that look like stretched-out Buckyballs. Nanotubes are strong and flexible and can be made in almost unlimited lengths. As happened 10 years ago, the sense of great possibilities for this new molecule is powerful. Richard Smalley told the *New York Times*, "[Buckyballs] are not nearly as interesting as the tubes. Bucky's biggest legacy has been to show us about the tubes."

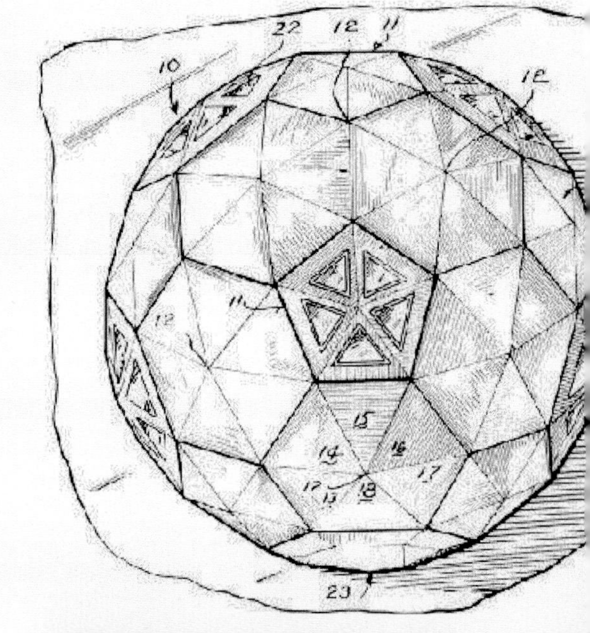

precedent in history." And a spherical building would enclose the maximum amount of space using the minimum amount of materials.

After working out the complicated mathematics involved, Fuller sketched out his first geodesic dome, made up of hundreds of interconnected triangles, each of which reinforced its neighbors through tension. In the summer of 1949, he and some students tried to build it, using aluminum slats from Venetian blinds. The slats were too flimsy, and the first dome collapsed into a twisted pile. Fuller was not discouraged. In 1950, he built a large dome in Montreal, and in 1953 he constructed a dome 50 feet in diameter at the Ford Motor Company's headquarters.

Thousands of people saw the Ford dome, and soon universities, businesses, and individuals were clamoring for a dome of their own. One of Fuller's biggest customers was the United States Air Force, which was setting up the Distant Early Warning system, a network of radar installations near the Arctic Circle that watched for an attack from the Soviet Union. Conditions in the Arctic were horrible, and some kind of shelter for the radar was needed. In response, Fuller designed the radome, a geodesic dome made from fiberglass struts that would not interfere with radar waves. The dome itself was perfect for the conditions: It could withstand very strong winds, and its nonmetallic surface never iced up.

The simplicity, beauty, strength, and novelty of the geodesic dome captured the public's attention, and Fuller himself soon became a symbol of a future based on new, nondestructive technologies. Some 300,000 geodesic domes have been built, serving as restaurants, factories, and even private houses. The most famous dome, at Disney's Epcot Center, attracts millions of visitors every year.

Today, some of Fuller's lessons on how to build using tension are being used in the design of new buildings. But people have been slow to adopt many of his ideas. Fuller always tried to think 50 years into the future. He knew that his ideas—whether on new ways to build things or what people should eat or the basic geometry of the universe—were too radical to be realized during his own lifetime. Yet he succeeded in inspiring a generation of architects, designers, and others to begin thinking in new directions, examining how to build, how to live, and how to make the world a better place.

The Wichita House, a prototype of Fuller's Dymaxion House, was built in 1947. The round structure's walls and roof were supported by cables attached to a central mast, since steel can support more weight suspended from it than placed on top of it. The house was reconstructed at the Henry Ford Museum & Greenfield Village in 2001.

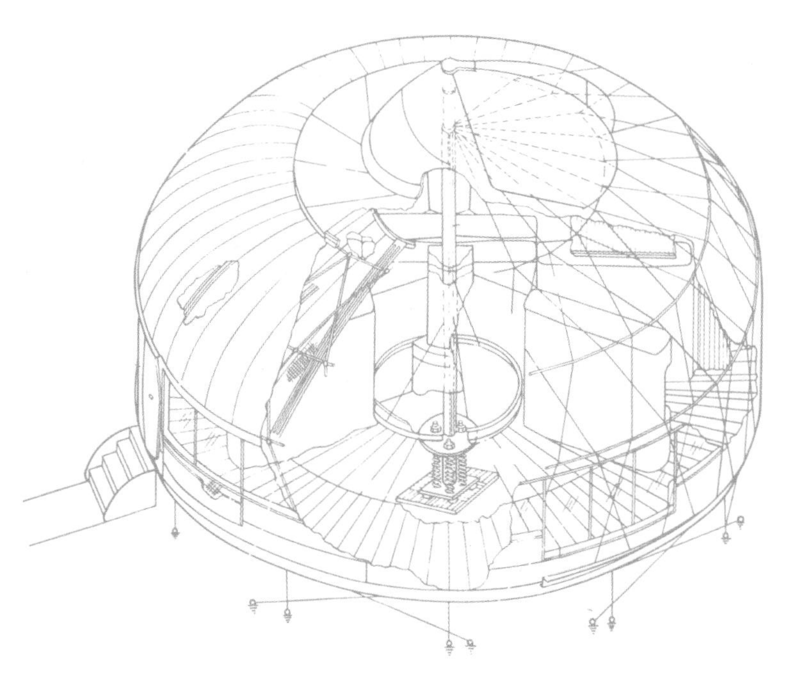

ASHOK GADGIL

One of the world's most serious health problems is the lack of safe drinking water in developing countries. By some estimates, up to two billion people cannot get clean water. Diarrhea, cholera, hepatitis, and other diseases caused by contaminated water kill some five million people a year, including three million children. Many times that number get sick, and the growth of 60 million children is stunted by recurring diarrhea and other illnesses. Because diarrhea kills by dehydration, the lack of clean water only accelerates the process.

Although the problem is well known, it has not been easy to battle. Water becomes contaminated in many different ways, including inadequate water systems, flooding, and a lack of proper sanitation. Most solutions (such as large filtration plants or new water systems) are far too expensive or complicated for poor nations to implement. Because the problem exists primarily in poor countries, there seems to be little money to be made in fixing it, and most scientists concentrate on more profitable challenges.

In December 1992, an outbreak of a new and particularly dangerous strain of cholera began in southeastern India. Dubbed "Bengal cholera," within months it had spread throughout India and into neighboring countries, killing up to 10,000 people. The tragedy inspired Ashok Gadgil, a scientist working at the Lawrence Berkeley National Laboratory in California, to look for a new way to purify drinking water. Using science no more complex than the ultraviolet light emitted by an unshielded fluorescent lamp, he built a simple, effective, and most important, inexpensive water disinfection system. Dozens of these systems are now installed around the world, and many more will be in use soon.

Born in Bombay, India, in 1950, Gadgil was interested in science from a very early age. "I finished reading all the high school books on general science by the time I was in the fourth grade," he remembers. His parents indulged his interests by buying him chemistry and mechanics sets, books, and electrical components. "One great thing my parents did was bring home *Popular Science*, *Popular Mechanics,* and *Scientific American*," Gadgil says. "So even

A Bhupalpur villager uses a hand pump to feed water into Gadgil's purification system, housed inside the cement structure.

though I was in Bombay, I had access to the kind of excitement that these magazines conveyed." He and some friends also formed a small science club to perform their own investigations—they worked to build better gliders, tracked individual ants in their colonies, and even tried to collect methane gas from rotting biomass in stagnant pools left by the yearly monsoons.

Although Gadgil's parents wanted him to become a doctor, he thought it would be "extremely boring to be doing the same thing all your life." Instead, he chose to pursue physics. He received his bachelor's and master's degrees in physics in India and then, in 1973, came to the University of California at Berkeley to continue his studies.

At Berkeley, Gadgil continued to figure out how best to use his physics knowledge. But he also became aware of things outside of the classroom. "Two things really struck me while I was in graduate school," he says. "One was the high level of affluence in the United States. The second thing was getting a sense of how important it is to have science applied to address problems that are societally significant. They may not be scientifically significant, because the basic science is well understood, but nobody had bothered to take the final step to convert that into a useful technology."

"In the 1970s, when I came to Berkeley, the important thing was relevance," Gadgil notes. "Whatever you are doing, is it meaningful, is it of value?" He began to study solar energy as it applied to buildings—a discipline that had clear applications in the real world, since the United States was facing an energy crisis. After receiving his Ph.D., Gadgil worked for a while at the Lawrence Berkeley National Laboratory, then spent about eight months in Paris collaborating with scientists looking for a solar solution to heating buildings in the Peruvian mountains.

In 1983, Gadgil and his wife Anjali, who is also Indian, decided to return to their home country to try to give something back to their community. After they arrived, he says, "it was clear that the stuff I learned in the United States would not be the right thing to do in India. The problems facing the Indians, as a society or as individuals, are not even on the radar screen here." Rather, he wanted to bring to India the approach he had learned in Berkeley and Paris, to "develop the right research agenda, identify the right research problems, and roll up your sleeves and go do something about it."

For the next five years, he worked with a fledgling research institution to address some basic problems in Indian society, including how to heat water and houses, and whether the country should adopt time zones in order to conserve electricity. (The answer was no; using efficient fluorescent bulbs instead of incandescents would save far more power.) In 1988, frustrated by the shortcomings of India's bureaucracy and its higher education system, he returned to the Lawrence Berkeley Lab to work on problems of energy efficiency.

In 1997, one of Gadgil's systems was installed at this South African hospice for HIV-positive children.

Gadgil remained involved with science back home, though, occasionally sending papers and other information to his colleagues there. One of India's problems that haunted him was that of contaminated drinking water. When Gadgil was living in India in the 1980s, he had witnessed a water-spread outbreak of hepatitis that affected at least 700 people. So in the early 1990s, he began sending information back to his colleagues in India about ultraviolet disinfection, which had some potential for cleaning up the country's water.

There are three basic ways to kill bacteria and viruses in drinking water: chlorination, boiling, and ultraviolet light. Chlorination depends on large-scale treatment plants, and thus is impractical in a country with a poor water infrastructure. Boiling water requires an impractically large amount of fuel, even more than is needed to cook food. And most of the energy used to disinfect water through boiling is wasted; all of the water needs to be boiled to kill the tiny amount of contaminants in it.

Ultraviolet (UV) light seemed like a great alternative to these methods. A certain wavelength of UV light has a profound affect on bacteria and viruses. When used for a short period of time—about 12 seconds—UV light damages their DNA, preventing them from reproducing or even making the enzymes that keep them alive. They die soon after this exposure—and, since they can't reproduce, even if live bacteria or viruses are ingested by a person, no illness will occur.

In 1993, after the Bengal cholera outbreak caused the death of more than 10,000 people in India and neighboring countries, Gadgil grew frustrated with his colleagues, who were not pursuing UV disinfection. That summer, he and a graduate student investigated the effectiveness of UV light and whether it was economically feasible. "We were completely amazed," he says. "Using the simplest engineering, we could disinfect water for half a cent per ton. That's shockingly cheap. You could disinfect one person's drinking supply for a full year for a couple of cents."

From his experiences in India, Gadgil knew that any purification system would have to require little maintenance and not take for granted basic technologies like electricity and water pressure. The system he and his student built, later named UV Waterworks, is remarkably simple. In a compact, enclosed box, a UV lamp is suspended above a shallow pan. Water runs into the pan by the force of gravity, where it is exposed to the UV light, then into a holding tank. The only power that is needed is about 40 watts to power the light; this can come from a car battery. The system can disinfect four gallons of water a minute, killing 99.999 percent of bacteria and viruses. Run continuously, this produces enough clean water to serve more than 1,000 people.

Water Health International, the company founded to bring the technology to market, now makes several different versions of Gadgil's disinfection system, for small and large applications, for emergency use, and for locations that also need to filter out silt and other large contaminants. Prices start at about $1,500. The company also offers solar cells and small windmills that can power the UV lamp.

UV Waterworks systems have been used in India, South Africa, the Philippines, Honduras, and other countries. Since 1998, the Mexican government has installed about 100 in Guerrero, a state on the Pacific in southwestern Mexico. The results have been very positive. In the summer of 2000, Gadgil reports, people from Water Health International returned with stories and data showing a dramatic decline in the incidence of diarrhea among children and adults. And preventing deaths and illness are just the most visible effects of purifying water—it also protects children from stunted physical and mental growth. "This is the kind of story that really makes my day, really makes me happy," Gadgil continues. "It makes me feel good when I get up in the morning."

Run continuously, Gadgil's UV Waterworks device can produce enough clean water to serve more than 1,000 people.

Stanford Ovshinsky

Stanford Ovshinsky built his career and several industries out of two basic ideas. The first was his concept for a completely new class of materials—disordered or "amorphous" materials. The second was a lifelong motivation: "I'd like to be able to make a better world." For more than 40 years, he has combined these two ideas in the pursuit of new kinds of electronics, batteries, and power sources that are more efficient, more powerful, and less polluting. Among the inventions that have come out of this quest are highly efficient batteries and hydrogen fuel cells to power automobiles, rewritable optical storage for computers, and solar cells that can be manufactured cheaply and in large quantities. His work, especially the batteries that powered General Motors' first production electric vehicle, the EV-1, led *Time* magazine to declare him a "Hero for the Planet."

It may be surprising that Ovshinsky, who holds almost 200 patents, is self-taught; he learned a trade during high school, and never pursued college. "I was not very much interested in school," he says. "But I was interested in learning." There were plenty of places for Ovshinsky to indulge his curiosity in Akron, Ohio, where he was born in 1922. He spent a lot of time around Akron's machine shops and foundries, and also read about all kinds of science, from anthropology and archaeology to astronomy. "At our local library, which was a small room in the poorer section of town," he recalls, "the librarians were only supposed to give out two books to every child, and only in their age bracket. They permitted me to take out as many as I wanted, of any age level."

During the depression, Ovshinsky attended high school during the day and trade school at night. After graduation, he opened his own machine shop, where he eventually began making machines that automated industrial processes. (The word *automation* hadn't been coined yet; Ovshinsky tried to get the term *automatation* to stick, although it didn't.) In the late 1940s, he brought his machines to Detroit and worked with the Big Three automakers—Chrysler, Ford, and General Motors—to improve production at their factories. In 1947 he created his first successful invention, an automatic lathe that he later sold to a large industrial company.

At Energy Conversion Devices' first laboratory, Stanford and Iris Ovshinsky discuss using solar energy to "disassociate" water, freeing hydrogen atoms.

The Ovshinskys founded Energy Conversion Devices in 1960 to further explore the amorphous materials he had been investigating and, they hoped, to create products that could better society.

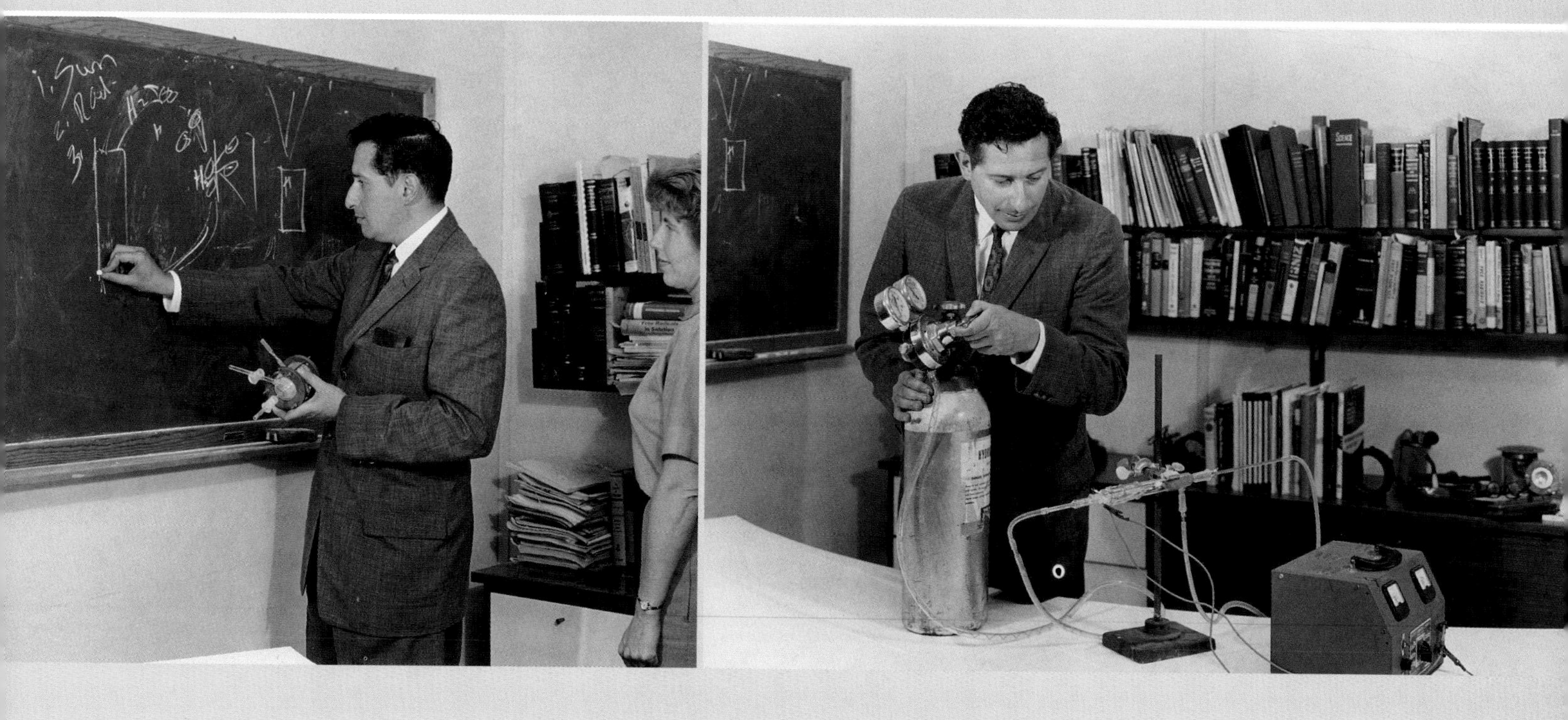

It was Ovshinsky's interest in automating machines that led him, in a roundabout way, to the science of amorphous materials. His machines could make some simple decisions as they went about their work, and Ovshinsky realized that this was a form of intelligence. But it was also clear to him that this wasn't *really* intelligence. So Ovshinsky asked, "What is the physical basis of intelligence?" He corresponded with several leading neurophysiologists about the question. "Some of them said that a nerve cell was like a crystalline material," he remembers. "One of them said that he thought that the neuron was an empty bag. And I didn't think that."

If a neuron was empty, Ovshinsky thought, then perhaps the outside of the neuron, its surface, was the important part. And while many researchers proposed that neurons worked like transistors—logic gates made out of crystalline silicon—he noticed that the neuron surface was disordered, noncrystalline. With the help of his wife, Iris, a biology and biochemistry Ph.D.,

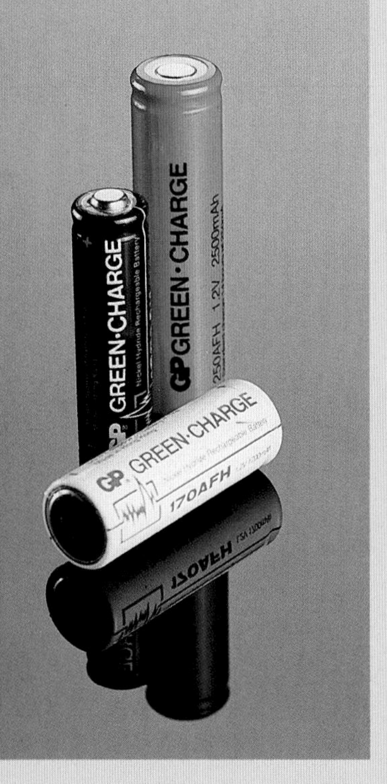

One of the major focuses of Stan Ovshinsky's career has been batteries. Licensed versions of his rechargeable nickel-metal-hydride (NiMH) batteries (above) are used in millions of laptops and other portable electronics. Four packs of NiMH batteries power this electric scooter (top), giving it a range of up to 70 miles.

Ovshinsky designed a model of a nerve cell based on this idea of a disordered surface. "It acted like a nerve cell should act," he says. "It had memory, it had the equivalent of synaptic connections."

This proved to Ovshinsky that disordered materials had interesting characteristics. Yet "I couldn't find any literature on disorder," he notes. "At that time amorphous materials were called *schmutz* materials." On January 1, 1960, he and Iris founded Energy Conversion Devices (ECD) to explore the field and, he hoped, "to have science and technology solve serious societal problems."

The most surprising and exciting property of disordered materials was that they could act as semiconductors. Until then, people thought that only crystalline solids, like the silicon that makes up computer chips, could act as semiconductors. Silicon does the job very well. But the process of manufacturing silicon chips is difficult, requiring "clean rooms" in factories that now cost upward of $1 billion, and the size of the chips is limited by the small size of a crystal. Disordered materials, as their name implies, do not have any overall structure. But their molecular structure—they are made from cesium, tellurium, or related elements—allows them to shift, in small quantities, to a more ordered state, when energy such as light or electrons hits them. This shift, called a "phase change," lets the material act like a conductive metal.

Ovshinsky's discovery that these materials could be shifted into working as a semiconductor opened up all kinds of possibilities, such as using them for computer memory. Because the phase change is a physical process—the molecules actually move into an ordered state—data can be stored even in the absence of electricity. This quality led to the invention of "flash" memory, often used in laptops and other electronic devices that need to store information when they're switched off, as well as rewritable CDs and DVDs.

In the early 1980s, Ovshinsky began applying his amorphous material science to solar panels, called photovoltaic (PV) panels. Photovoltaics had until then been made using crystalline materials like silicon. As with computer chips, the manufacturing process was expensive and demanding. In addition, these PV panels were thick and heavy, so they had to be mounted on a building or some other strong structure.

Although the PV cell Ovshinsky developed using disordered materials only produced about 50 to 60 percent of the power that conventional PV did, it had some significant advantages. The amorphous semiconductor that made them work was a thin film, rather than a thick crystalline material, so the PV cells were much lighter and more flexible. Most important, they allowed Ovshinsky to create a new PV manufacturing process. In 1983, he patented a device much like a large printing press. A long roll of thin, flexible metal is fed through the machine, which deposits the thin film of disordered materials. The most recent generation of the machine can create 1/2-mile-long rolls of PV cells up to 18 inches wide; the next version will accommodate 1 1/2-mile-long rolls.

Such large-scale production has drastically lowered the cost of PV cells, to about $1 per watt, which is comparable with power from oil, gas, or coal. In addition, the thin, lightweight material has freed designers to create new kinds of energy-producing components. The most interesting of these is the PV shingle, which not only looks like a slightly oversized version of a regular roof shingle, it works like one. "What I've done with PV is make it prosaic, bring it back down to earth," Ovshinsky says. "The shingles aren't on top of the roof, they are the roof."

To Ovshinsky, this is an important achievement. The mission of his and Iris's company is still to change the world for the better, and he believes you can't do that with niche products, as PV cells have been for decades. PV shingles are easily understandable by people who might not be interested in solar energy; they also save space, since there are already roofs all around the world ready for them.

Ovshinsky's company is also involved with another kind of energy technology: hydrogen fuel cells, which may turn out to be even more valuable than solar cells. A fuel cell uses a reaction similar to that of a battery to create electricity from hydrogen; many car makers are betting that it is the next big fuel technology. There is no pollution—the only by-product is water—and hydrogen is abundant.

Although it is an inexpensive fuel source, hydrogen is difficult to store and manage. As a gas, it has to be stored under great pressure; as a liquid, it has to be kept at −253° Celsius. Ovshinsky devised a better way to store it: in metal. Hydrogen can be pulled out of or pushed into a metal matrix, which is compact and safe and stores more hydrogen per liter than the liquid or gas forms of the element. Making hydrogen easier to use will accelerate the development of fuel-cell–powered cars.

Even though humans only recently learned how to convert sunlight into usable energy, Ovshinsky promises that hydrogen power could be even more revolutionary. "It's a complete decoupling from the earth's resources," he says. "It's pollution-free. You don't have to fight wars over oil." Still going strong at the age of 78, Ovshinsky sees his latest technological contribution as just part of the goals he and his wife set for ECD in 1960. "We're doing what we said we were going to do," he says. "What better way is there to express your ideals than to make the world better and safer and healthier?"

Ovshinsky discovered that amorphous materials could act like semiconductors; one application of this was the creation of thin, flexible solar power panels. Making the panels in large rolls lowers the cost of their power to about $1 per watt.

JOHN TODD

John Todd is dedicated to the idea that biology should be the model for all design. By enlisting nature's own processes, he has created new systems for farming, aquaculture, treating waste, and purifying water, among other things. These systems, which he calls Living Machines, have the potential to shift technology—and by extension the world—to a more sustainable path.

This application of biology and ecology to human systems is fitting, for Todd grew up surrounded by both nature and industry. Born in 1939, he was raised on the shores of Hamilton Bay, a center for Canadian steel and shipping. It was "a world of ships and slag and burning ovens and fires and ice," Todd says. Although his father was an executive for 3M, he was also passionate about the outdoors. "All he talked about at home was the woods and the farms and the forest and canoes and kayaks," Todd remembers. So while the local environment had been severely damaged, he was still able to develop a deep connection with nature. "We had marshes right next to our house and streams and I could follow them right up to their headwaters," he says. "It was a very mysterious world for a kid."

Todd had little interest in academics as a youth, preferring pursuits such as sailing and skiing. Two events helped change his focus. The first was his becoming depressed as the local environment was further damaged. "Some of the woods I would walk through to school were being cut down," he explains. "Orchards and market farms were being paved over to be subdivisions. Even my beloved clay pits were under siege. From all around, from the water to the hills behind, I felt threatened." Then, when Todd was 13, his father gave him a series of books written by the novelist Louis Bromfield, describing his effort to restore an agricultural region of Ohio.

Bromfield had returned from Europe after World War II to find the area depopulated and eroding. "His dream was to bring it back to life," Todd says. "He re-created an agricultural community with the holistic knowledge he had brought from [rural France]. In the midst of destruction, here was this marvelous tale of hope." Todd began working on farms and decided to study agriculture at McGill University in Montreal, Canada. What he was taught at

Todd's first attempt at using living things to clean the environment took place in Cape Cod—he set up 21 translucent tanks to cleanse sewage being dumped into an open pit.

college, though, came as a shock. The agriculture program focused on chemical and mechanical farming, not sustainable or traditional practices. "We were told that most of us would be salesmen for the drug or chemical companies," he remembers.

Dispirited by the idea of institutional farming, Todd became interested in the disciplines of his favorite professors: ecology and biology. After graduating from McGill, he worked for several years as an environmental consultant, surveying damage from factories across America and Canada. Even in these affected environments, though, he found some reasons for hope. "I could see points where there was a recovery," he says. "These were my first inklings of the possibility of the regenerative forces of nature."

In graduate school, first at McGill and then at the University of Michigan, Todd studied oceanography and fisheries. After earning an award for the

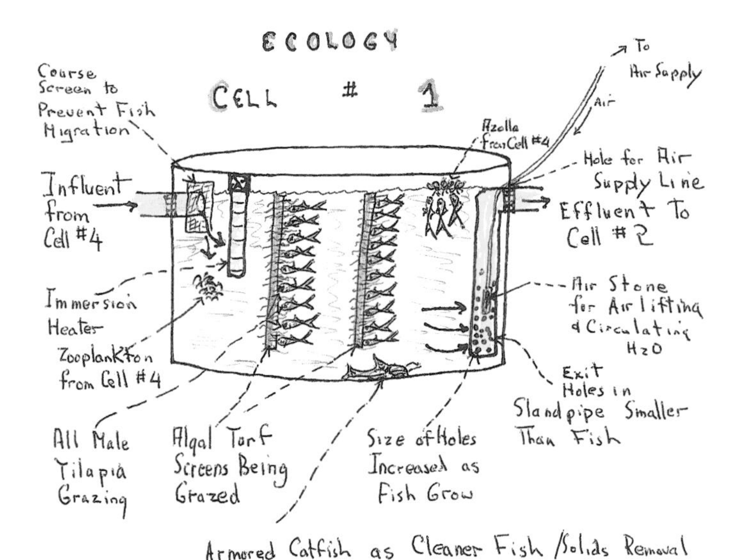

outstanding doctorate at Michigan in 1968, he decided to take a position at San Diego State University teaching ethology (animal behavior) so he could conduct research in the warm waters of nearby Mexico. After a couple of years of "doom watch" research—studying what's wrong with the environment—Todd felt he'd reached a dead end. With the help of his wife, Nancy, he began to look at more positive possibilities for science. "I began to have a conversation around how knowledge could be reintegrated to be reused in an earth-healing way," he says.

He, Nancy, and their colleague William McLarney began to pursue what they called "new alchemy," the science and practice of applied ecology. In a 1971 paper, Todd laid out a plan for a new science of "earth stewardship." The three of them also founded the New Alchemy Institute in Woods Hole, Massachusetts, where Todd had taken an oceanographic research job. Today, he realizes that their goals seem innocent and perhaps pompous. "Our slogan was, 'To restore the land, protect the seas, and inform the Earth's stewards,'" he recalls. "That's a lot for three people."

During the 1970s, New Alchemy's work became more concrete, as Todd and others began to establish guidelines for

John Todd's Living Machines use natural processes—from bottom-feeding fish to photosynthesis—to remove contaminants from water.

designing systems from a biological or ecological starting point. In the mid-1980s, Todd decided that one of the biggest threats to humans was water pollution. Two of his closest friends had recently died of cancer, and he noticed that a neighboring Cape Cod town was dumping its sewage and other waste in an open pit just 25 feet above the drinking water table. Looking at that pit, Todd says, he "made the connection between the environment and this epidemic that permeates our lives."

Methods for treating this kind of waste already existed, but they were chemical- and energy-intensive and expensive, and not being used at all in this case. From his own experience, Todd thought he could build a treatment system that was less expensive and based on natural processes. In the 1970s, he had built seminatural habitats to conduct research on fish from the Pacific Ocean and the Gulf of Mexico, fish that were notoriously hard to breed in captivity. In essence, he re-created their natural environment in a greenhouse. "I was asking plants and animals to play the roles of the filters and all the other things you would find in an aquarium," he explains. Todd succeeded in breeding the fish, but he wound up more interested in the system he had built. "It was clear in my mind," he continues, "that living systems, whether plant or animal, can be a substitute for energy, hardware, and manufactured chemicals. There is an analog that we can find in the natural world to carry out functions that we currently carry out with brute force."

In 1984, Todd applied these ideas to a test system to treat the waste in that open pit. He built an artificial marsh and 21 connected clear tanks, each containing various organisms, from algae and marsh plants to invertebrates and small fish. As the water made its way through these tanks the living creatures in each broke down, ate, or otherwise processed the pollutants. The process was sped up by the tanks' transparency—bright sunlight accelerated the photosynthesis that fueled many of the organisms. The cleansing process took 12 days, at the end of which 100 percent of nearly all the major pollutants had been removed.

Two companies have been founded to refine and market Todd's water treatment system. To date, it has been installed in Florida, Wisconsin, Vermont, and Nevada, among other places, to treat waste from factories, nature centers, and schools. These Living Machines work much as the first test system did. Waste water is first directed into underground tanks, where organic matter and solids are broken down by bacteria. Then, the water moves through a series of open-air tanks full of plants,

At the San Francisco Museum of Modern Art, a small Living Machine supported a species of African fish that is extinct in the wild.

Today, a Living Machine in Providence, Rhode Island, treats millions of gallons of sewage every year.

Another Living Machine was installed in Frederick, Maryland, as a demonstration of an ecologically engineered landscape.

plankton, snails, and fish, which each play a role in digesting and breaking down more of the waste. Remaining solids are collected from the water, which then moves through a natural filter such as lava rock. In the end, the water is clean enough to be reused or released into local waterways.

Todd continues to work on other designs that harness natural processes. He is currently developing a method for "compact farming" that would let people raise food in relatively small spaces. "In a sense I'm trying to reinvent a chunk of agriculture," he says. If his new method of farming is adopted, it could bring the production of food closer to urban areas and eventually open up more space for nonindustrial and nonfarming use.

Todd is also working with architect Maya Lin on the restoration of a chapel in Purchase, New York. A glass structure around the ruined church would stabilize it, and a Living Machine would be installed inside to clean the water of a nearby creek. Water "would flow up into the tower and then through the building in a very formal way," he explains, "through a Living Machine, step by step, right down the middle of the building. The purified water then heads back to the stream." This marriage of nature and aesthetics encapsulates many of Todd's ecological beliefs and designs. With the Living Machine's clear tanks occupying the chapel, he adds, "the connection between other forms of life and our own lives will be rendered more visible."

"Our slogan was, 'To restore the land, protect the seas, and inform the Earth's stewards,'" he recalls. "That's a lot for three people."

COMPUTING AND TELECOMMUNICATIONS

INTRODUCTION

This group of innovative American minds works in what is perhaps the key field of endeavor when it comes to causing change in the modern world: information and communications technology. Throughout history, the greatest innovative surges seem to come after a major advance in the ability to generate, store, or distribute information. Developments in communications technology make it easier for ideas to come together in new ways and generate novelty. Innovation generated in this way then expands by a ripple effect, spreading out from the original idea to bring social change.

Witness the ripple effect of one side-product of the printing press: printed maps. In the early seventeenth century these maps, easily updated and reprinted with the latest navigation and exploratory data from ships' logs, made it safer for captains to go looking for profitable cargoes in the East. The first Dutch trading companies to use the new maps made it a legal requirement for returning crews to report all newly discovered navigational information to the company printing-house, so that the data could be included in the next edition of charts and atlases. As a result, by the second decade of the century the Dutch were bringing goods back from the East and re-exporting them all around Europe, sometimes making 600 percent profit in the process. The idea soon caught on, and demand for cash to invest in the new ventures triggered the establishment of new land registers to give formal title to land (the prime contemporary source of wealth) so as to borrow money against it, through another new entity set up for the purpose: the mortgage company. Insurance companies were then organized to reduce the financial risk of shipwreck, and joint-stock companies were formed to reduce the risk of failure. Joint-stock companies in turn brought the emergence of the first stock market and a national bank was set up to organize extra funding through its new credit agencies. The whole enterprise was made secure with another new invention: the business contract, whose philosophy would find its way into politics with the citizen-state relationship embodied in the new Constitution of the United States.

The information/communication-technology work of this group of inventors—including the mouse, the walkie-talkie, computerized telephone switching systems, microchips, programming, computer games, and the personal computer—will help to generate perhaps the greatest ripple effect so far. By the end of the first decade of this century the communication and information industry can expect to see virtually unlimited bandwidth, handheld semi-intelligent communications devices, and supercomputers capable of a thousand trillion operations per second. What this technology will do can only be guessed at. Continuing innovation in this field will almost certainly make America the first fully virtual social and economic structure. All citizens, anywhere, will be able to enjoy the benefits of a new kind of existence, wherever they find themselves: in education, jobs, acts of self-expression, with access to anything they want to know, wherever they are, at any time of the night and day. Electronic avatars will represent them in social, political, and business decisions, making these future Americans members of the first truly informed and fully enfranchised society in history.

Perhaps the greatest moment of discontinuity that our final group of inventors is helping to facilitate, however, will come when the new technology changes the nature of innovation itself. Ever since the first flint tool, the constraints imposed by technological limitations have shaped what contemporary society could do and, above all, who could do it. Access to knowledge and the means of expression was limited to a tiny, literate minority. It is their work alone that tells us what happened in history. We know little or nothing of the vast, silent, illiterate majority. This state of affairs has lasted up to the present time and has generated what might be termed a "culture of scarcity," in which the few have always ordered and the many have always obeyed. We have lived for millennia in a culture in which talent and intelligence and innovative capabilities have come to be regarded as exceptional, because in the country of the blind the one-eyed man is king.

We have no means of knowing if those individuals whom we now regard as history's greats—Newton, Michelangelo, Franklin, Galileo, and all the others—were indeed the best. There may have been thousands of others even more talented, but, being illiterate, they were out of the loop. However, the ones we know about certainly showed what humans are capable of, given the tools and the opportunity. This is what the information and communications technology now being developed by American inventors is about to offer, to an unprecedented number of human beings, all over the planet, each of whom has a brain of the same size and complexity as Einstein's.

—JB

NOLAN **BUSHNELL**

In just three decades, video games have taken a prominent and permanent place in American culture. They have more than a hundred million fans, have spawned dozens of magazines, and have even captured the attention (and disapproval) of the U.S. Senate. The demands made by games have been responsible for many improvements in computer technology, advances that have led to higher graphics and speed capabilities in today's personal computers.

The more than $6 billion Americans now spend on video games every year started with the first quarter dropped into Computer Space in 1971. That game—a small computer hooked up to a black-and-white TV, housed in a futuristic-looking plastic case—was the creation of Nolan Bushnell, a young engineer from Utah. Bushnell went on to found Atari, whose products, from Pong to Football to the Atari 2600, brought video games into every arcade and millions of homes. And while Computer Space was based on the already-classic computer spaceship battle game called Spacewar, it was Bushnell's genius to see the potential games had beyond the computer lab.

"I've always been a tinkerer," Bushnell says. As a teenager, he was one of the youngest ham radio operators in the country, and he did science experiments in his garage. At least one of those went awry: A liquid-fuel rocket attached to a roller skate crashed into the back of his garage, igniting a bright but short-lived fireball. Rockets were a side interest, though. "I loved electronics from an early age," Bushnell remembers, "But I was also always a game player." He was a tournament chess player, and a fan of the Chinese board game Go. (Atari is a Japanese word announced when a Go player has almost captured an opponent.) He also learned about business when he was young. After his father died, Bushnell took over the family's concrete business. He was just 15.

Bushnell discovered computer games in the early 1960s while studying electrical engineering at the University of Utah. The school's computer had a copy of Spacewar, the seminal game created at MIT by Steve Russell. Two starships circling a sun battled each other and the star's gravity, firing "torpedoes" and occasionally resorting to "hyperspace" to jump out of, and

In 1977, Atari released the Video Computer System, also known as the 2600. More than 25 million of the cartridge-based machines were sold, ushering in the first golden age of home video games.

The first successful video arcade game was Atari's Pong, which debuted in 1972. The game, a simple electronic tennis match, had just one line of instructions: "Avoid Missing Ball for High Score."

THE PREHISTORY OF VIDEO GAMES

Late eighteenth century—Bagatelle, a miniature tabletop game related to billiards, becomes popular.

1871—A small version of bagatelle with a spring-loaded plunger becomes popular; it is the ancestor of the pinball machine.

1905—The first nickelodeon opens, in Pittsburgh. Customers can see a short movie by dropping a nickel in a machine's slot.

1931—The first coin-operated pinball machine is built.

1947—The pinball flipper is invented, ushering in the Golden Age of Pinball. The game is banned in New York City the following year.

1958—Willy Higinbotham, an engineer at Brookhaven National Laboratory on Long Island, New York, uses an analog computer to make a "ball" bounce back and forth across the screen of a small oscilloscope. He adds a "net" and the first computer tennis game is born. Visitors to the lab's open house line up for hours to play.

sometimes into, trouble. (Hyperspace would reappear in 1978 in Atari's Asteroids game.) Spacewar was immensely popular among computer scientists and students, and was soon being played on computers around the world. The visionary computer designer Alan Kay once declared, "The game of Spacewar blossoms spontaneously wherever there is a graphics display connected to a computer." Bushnell was hooked, and he would sneak into the computer lab late at night to play.

During his college summers, Bushnell managed the arcade at a local amusement park, where he saw how eager people were to play games. It was this experience that gave Bushnell his big idea. "It was very easy for me to see that if I could put a quarter slot on [Spacewar] and put it in the amusement park, it'd make money. It wasn't really a huge Aha!" he says. But Spacewar ran on huge, expensive computers. "You'd divide 25 cents into a $7 million computer, and you'd see, gee, this doesn't work," he continues. "But in the back of your mind, you'd say, If I can build it cheap enough it will work. And one day the costs came in such that I thought that it was possible."

By 1971, Bushnell had moved to Silicon Valley and had begun to work on his game. He took over one of his daughters' rooms and built from scratch the circuits that would run his own version of Spacewar, which he called Computer Space. The biggest technical challenge was the display. The computers that ran Spacewar used what were essentially adapted radar screens, each of which cost about $30,000. So Bushnell made circuits that would display graphics on an ordinary black-and-white television set. At his kitchen table, he sculpted a scale model of Computer Space's curvy, futuristic case out of clay. (The game looked so distinctive that it was used as a prop in the 1973 sci-fi movie *Soylent Green*.)

Computer Space was licensed to a company called Nutting Associates, which built about 1,500 of the machines. The game was a moderate success, but not a runaway hit; many people thought it was too complicated. Bushnell had a different complaint about its manufacturer. "Nutting Associates was an extremely incompetent company," he says. "I've often thought that that was a huge blessing, because it was clear that these guys were succeeding in spite of being really stupid. There's nothing like being around that to embolden yourself."

Bushnell began designing other games and he hired a staff of engineers. In 1972, Bally, a company that made pinball and slot machines, contracted him to make a video driving game. He gave the task to one of his new hires, Al Alcorn. But Alcorn didn't yet know the tricks of making a video game, so Bushnell gave him a smaller task: to make a game with a ball bouncing back and forth on the screen. (Bushnell had seen a similar game demonstrated that year as part of the Odyssey home video

1961–1962—At MIT, Steve Russell and others create Spacewar on the university's PDP-1 computer. Copies of Spacewar wind up at universities around the country, draining countless hours of expensive computer time and inspiring a generation of programmers. The first gaming controls, the ancestors of the joystick, are built by Spacewar-playing MIT students.

Mid-1960s—More games are written for the PDP-1 and other computers, including Lunar Lander, which simulated a moon landing; Hunt the Wumpus, a search-and-destroy mission; and ADVENT, the world's first adventure game.

1966–1967—Ralph Baer, a defense company engineer, builds a small computer and hooks it up to a TV set and makes two electronic dots chase each other across the screen. Within a year, he and some colleagues turn it into a system that plays Ping-Pong, football, and shooting games. When it's released as the Magnavox Odyssey in 1972, Baer's invention becomes the first home video game system.

Nolan Bushnell sold Atari in 1978; since then, he has been involved in business ranging from family restaurants to home robots to Internet gaming.

"The brain is something that if you exercise it you can be smarter. It turns out that games are that exercise."

game, designed by Ralph Baer and marketed by Magnavox.) "I defined this very simple game for Alcorn as a learning project," he explains. "I thought it was going to be a throwaway. It took him less than a week to get it partially running. And the thing was just incredibly fun."

Bushnell took a copy of Alcorn's game, named Pong after one of its noises, to Bally's headquarters in Chicago, hoping that they would buy it instead of the driving game. At the same time, Alcorn built a case for their other copy of Pong, complete with a 13-inch TV set and a slot for quarters. There was one sentence of instructions on the cabinet: "Avoid Missing Ball for High Score." Alcorn set Pong up at a bar in Sunnyvale, California.

In Chicago, Bally's turned Pong down. Back in California, the reaction was different. People lined up to feed quarters into Pong, and played it nonstop. The next day, the machine suddenly stopped working; Alcorn went to see what was wrong and discovered that the machine was too full of quarters—they'd spilled out of their container and shorted the game out. Pong, released by Atari rather than Bally's, became a hit and ushered in the first golden age of video games. Rich from Pong's success, the company designed dozens of successful games, first variations on Pong and later arcade hits like Atari Football, the driving games Night Driver and Sprint, and, in 1978, the best-selling Asteroids.

Bushnell also helped usher in a new era in Silicon Valley. Although the area (which extends from Palo Alto to San Jose) had long been a center for the electronics industry, most of the companies there were large and corporate. Atari was different. Bushnell always wore jeans, and he encouraged his engineers and technicians to do the same. His management style was not very rigid or hierarchical; as long as someone got his or her job done, almost anything went. These principles were proved in 1976, when Bushnell hired a young technician named Steve Jobs. The long-haired Jobs would often work barefoot, talked of going to India, and was abrasive to some of the other engineers. Bushnell gave Jobs the task of designing a game he had thought of, a new variation on Pong called Breakout. Jobs worked the night shift, and with lots of technical help from his friend Steve Wozniak, built Breakout on a very short schedule. The two would continue their collaboration that year by building and marketing the Apple computer.

In 1975, Atari introduced a version of Pong for the home, which quickly sold hundreds of thousands of copies and spawned dozens of imitators. Two years later, the company introduced the Video Computer System, also known as the Atari 2600. This was the first successful home system that used interchangeable cartridges, so players could enjoy dozens (and eventually hundreds) of different games. More than 25 million systems were sold, and the system-and-cartridge idea has been adopted by Nintendo, Sega, and Sony. "That's something I'm really proud of," Bushnell says. "I set up the economic model, the software/hardware link that's been the pattern ever since. There wasn't anything out there like it."

Bushnell sold Atari in 1978 and left the company soon after that. He has launched several businesses since then, including the Chuck E. Cheese chain of video-game pizza parlors and, in the 1980s, Axlon, which made small robots for the home. He has now returned to his video game roots, running a company called uWink that is introducing a line of arcade games that will be linked up over the Internet. Bushnell envisions game tournaments with thousands of participants linked up around the country or the world.

Almost thirty years after building that first arcade game, Bushnell remains excited about the promise of video games, as much for their effects on players as for their business opportunities. Just as Spacewar attracted him to the computer lab as a youth in Utah, today's games excite kids about technology, and even make them smarter. "I call them the training wheels of computer literacy," he says. "It's very clear that game playing grows dendrites. So people are smarter. The brain is something that if you exercise it you can be smarter. It turns out that games are that exercise."

DOUGLAS ENGELBART

When Douglas Engelbart took the stage at the San Francisco Civic Auditorium in December 1968 to demonstrate the NLS (oN Line System), a computer he and some colleagues at the Stanford Research Institute had built, only a handful of the 2,000 people gathered for the Fall Joint Computer Conference could have imagined how prescient the demo would prove. Looking a bit like a NASA Mission Control navigator, Engelbart sat at a computer console and spoke into a headset microphone while an image of his face was projected onto a screen behind him. He explained to the crowd that he was going to show them how his system worked, not just explain it. And he began. Overlaid on the image of his talking head was a computer display, the kind of text-only display that a generation of personal computer enthusiasts would come to know, but that until then had never been seen. He typed a few words on the blank screen. And then Engelbart began to manipulate the words he had typed. A strange object appeared on the screen: a small, moving black dot.

The little dot is what we now call a cursor. Engelbart was moving it with a peculiar device under his right hand, a chunky, square box with a single button, tethered to the machine with a cord. In the lab, he and his colleagues had called it a "mouse," after its tail-like cable, and the name stuck. With this mouse, Engelbart was able to select text, move it around, and otherwise manipulate it. The mouse was not simply a pointing device, though. It was a key element of the NLS, which in turn was a key element of Engelbart's larger ideas about "Human Augmentation"—making systems and tools that would help people work smarter and better. He has been pursuing those ideas since he had an epiphany about what to do with his life in 1950, while driving to his job as an engineer at the National Advisory Committee on Aeronautics (NACA, which became NASA) Ames Research Center in Mountain View, California.

It had taken Engelbart the first 25 years of his life to get to that moment of clarity. Born in Oregon in 1925, he was the grandson of Western pioneers and the son of a radio store owner and a mother who, he remembers, was "quite sensitive, and artistic." His father passed away when Engelbart was only nine, and the family's finances were tight. After high school, he studied electrical engineering at Oregon State University for two years, then signed up with the navy. This was in 1944, the thick of World War II, though on the day he shipped out of San Francisco, Japan surrendered.

In the navy, Engelbart served as a radar technician, which gave him direct experience with how information could be conveyed directly, electronically, on a screen. While he was stationed in the Philippines, he came across a copy of *Life* magazine that included Vannevar Bush's essay, "As We May Think," which described something like the personal computer, as well as something like the World Wide Web.

After the war, Engelbart finished his degree and moved down to the San Francisco Bay area, to work at Ames. Within a few years, his work there had become less than exciting, and he began to wonder what else he could, or should, be doing. The epiphany came to him the day after he asked his future wife, Ballard, to marry him. He was driving to work, thinking, and "I had this realization that I didn't have any more goals, and that getting married and living happily ever after was the last of my goals," he told a Stanford University interviewer in 1986. "I had to figure out a good set of professional goals." He calculated how many minutes of working life he had left in him—about five and a half million—and wondered how he could maximize the use of that time, for himself but also for mankind.

So Engelbart began looking at the various avenues he could take to do something beneficial and still make a living. (He had a family to support, and memories of his modest childhood weren't far from his mind.) He thought about going into economics, or education. But then he saw, suddenly and whole, the problem that he could help solve. Engelbart realized that everything had changed. The world was moving away from old modes of work, industry, and thought. Civilization had reached a point where the problems it was facing were growing increasingly complex, and at the same time they were growing more urgent. This complexity and urgency "had transcended what humans can cope with," he explained. "It suddenly flashed that if you

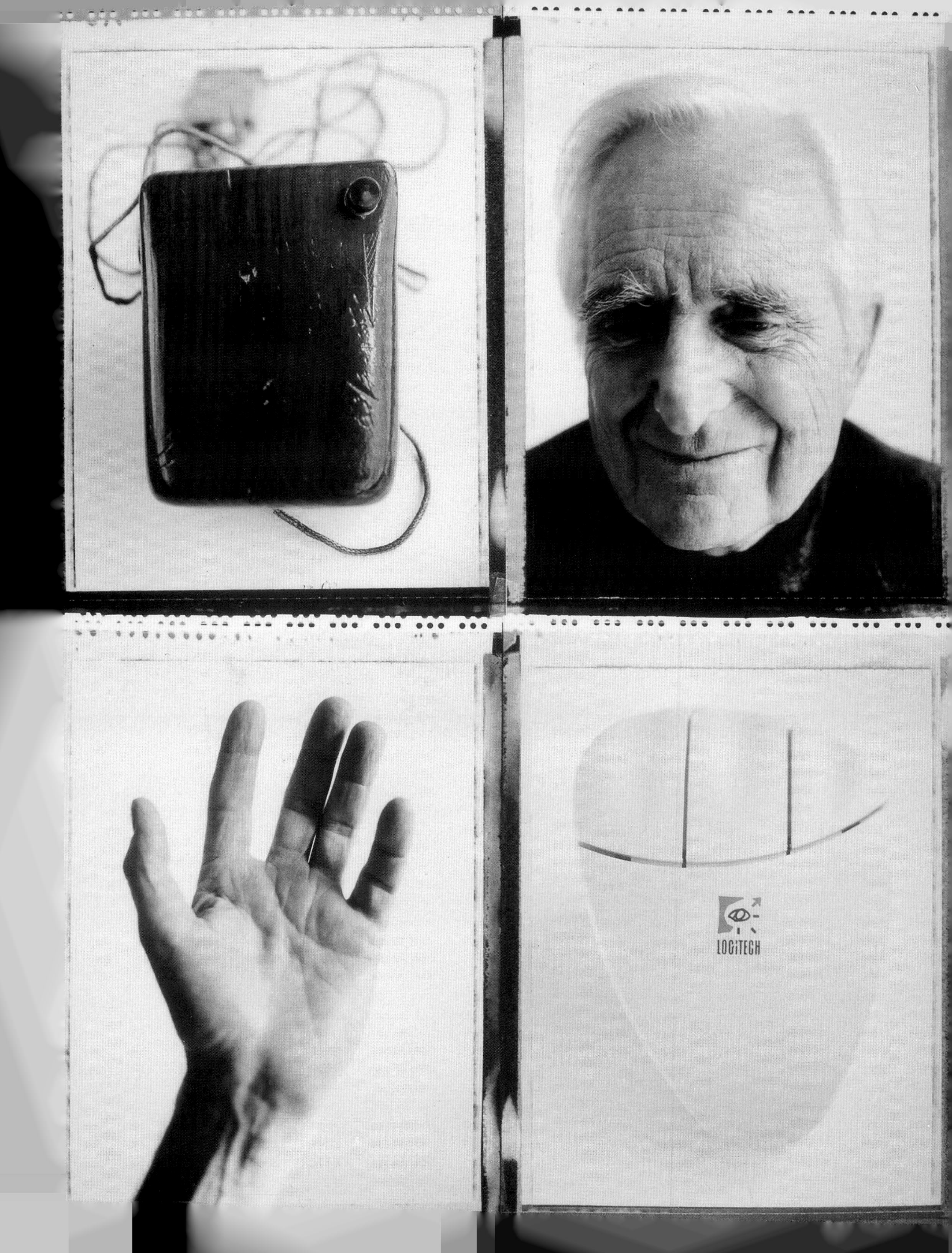
LOGITECH

could do something to improve human capability to deal with that, then you'd really contribute something basic."

What was needed to achieve this goal was a new way to work with information and for people to work with each other. "What makes us think that the old way of looking at a book page," he asks, "or of writing, that that's the way God meant us to do it?" Engelbart, who had read about digital computers and had experience with the display screens used in radar systems, quickly envisioned what such a system would be. "I think it was just within an hour that I had the image of sitting at a big screen with all kinds of symbols, new and different symbols," he remembers. "I also really got a clear picture that one's colleagues could be sitting in other rooms with similar workstations, tied to the same computer complex, and could be sharing and working and collaborating very closely."

It took 18 years to get from that vision to the demonstration in San Francisco. Within months, Engelbart left NACA to enter the electrical engineering Ph.D. program at the University of California at Berkeley, which offered the chance to study computers, although it did not have a computer itself. With his wild ideas about computers and interaction and collaboration, however, Engelbart was not cut out for academia, and after earning his Ph.D. he joined the Stanford Research Institute (SRI), an independent think tank in Menlo Park. In 1959, he began to outline the system he had envisioned a decade earlier, and in 1963 published a paper entitled "A Conceptual Framework for the Augmentation of Man's Intellect." The paper caught the eye of the U.S. Advanced Research Projects Agency (ARPA, now DARPA), and Engelbart was able to found the Augmentation Research Center at SRI.

His research center set out to create new tools for working with ideas and words. There were two main ideas Engelbart followed in developing such a tool, which would become the NLS. First was the creation of a new "writing machine" that would let someone revise and reorganize text and ideas on the fly. In essence, this writing machine was a word processor. But in his new way to get words down on paper, Engelbart saw a path to more advanced thinking: "If the tangle of thoughts represented by the draft becomes too complex," he wrote, "you can compile a reordered draft quickly. It would be practical for you to accommodate more complexity in the trails of thought you might build in search of the path that suits your needs." The second

The world's first computer mouse, built by Doug Engelbart in the mid-1960s. The simple wood box has a button on top and two wheels underneath, to track a user's hand movements.

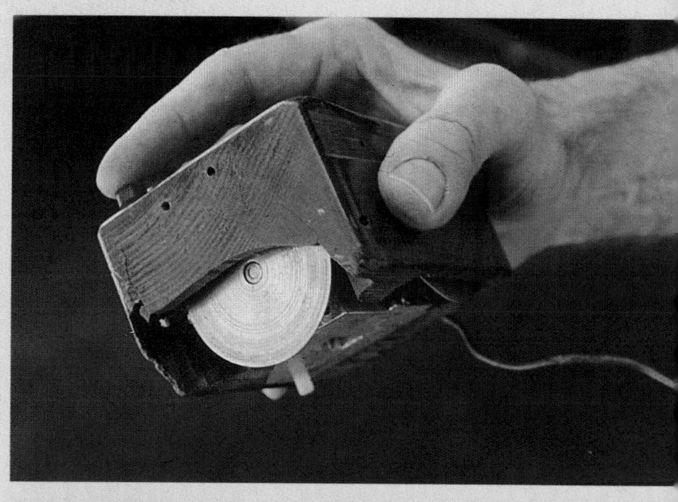

Engelbart's mouse and chord keyboard were small parts of his overall vision: to build computers that would let people work better, smarter, and in collaboration with each other.

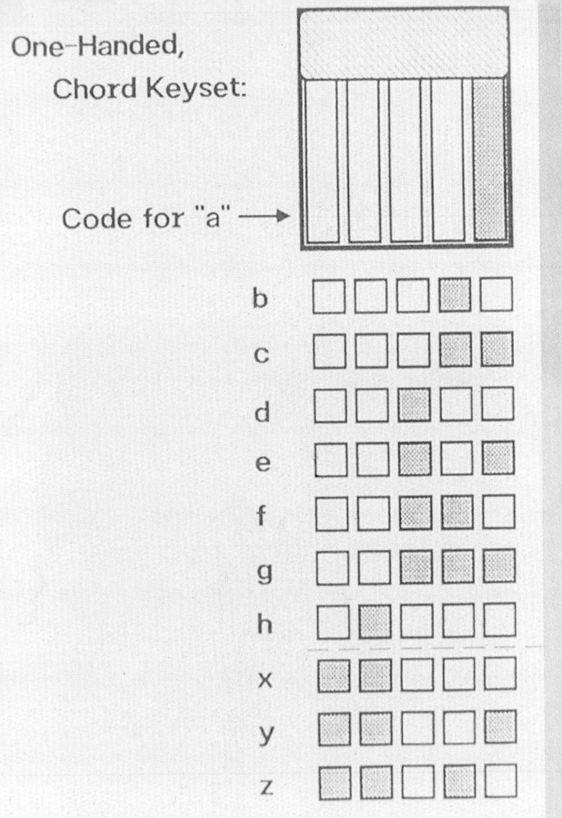

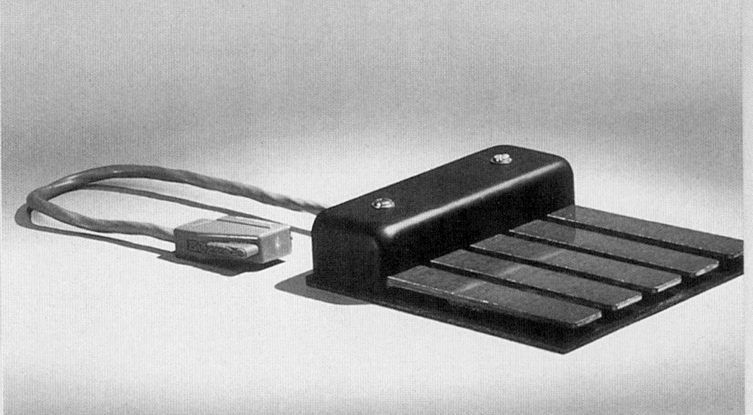

The computer system Engelbart built in the 1960s included not only a mouse but also a new kind of one-handed keyboard. A user typed by hitting a combination of its five keys, much like playing a piano chord.

AS WE MAY THINK

The essay that Douglas Engelbart happened upon in late 1945, in a Red Cross library in a native hut in the Philippines, has become a cornerstone of the Information Age, and the founding document of the World Wide Web. Published in the July 1945 issue of the *Atlantic Monthly* (and a few months later, in an abridged version, in *Life*), Vannevar Bush's "As We May Think" sketched out what the author saw as a looming problem: the inaccessibility of much of the information being produced. "We can enormously extend the record," Bush wrote, "yet even in its present bulk we can hardly consult it. This is a much larger matter than merely the extraction of data for the purposes of scientific research; it involves the entire process by which man profits by his inheritance of acquired knowledge."

Bush—a scientist, engineer, inventor, and, during World War II, the director of the U.S. Office of Scientific Research and Development—envisioned a technological solution, a new machine that would help humans navigate information in a way that mimicked the way we think. The mind works, he explained, by association. "With one item in its grasp, it snaps instantly to the next that is suggested by the association of thoughts, in accordance with some intricate web of trails carried by the cells of the brain."

The device Bush proposed was a kind of predigital personal computer that allowed an individual to store and retrieve "all his books, records, and communications." The desk-size Memex, as he named it, would have screens on which to display these records, a keyboard, and controls to move through one's store of information. This would have made for an interesting machine in

idea that guided him was that for people to work better, they had to be able to work together, to share their "trails of thought" and develop them collaboratively, in real time. The NLS allowed two or more users to work on the same document from different workstations, even from different locations. (At the 1968 San Francisco demo, a colleague in Menlo Park joined Engelbart on line, and on screen, to work on a single document.)

The early 1960s was an unlikely time to be building a computer intended for tasks such as writing and manipulating ideas. Since the digital computer's birth in the 1940s, its main work had revolved around crunching numbers. Power was more important than speed, and "ease of use" was unheard of. Computers required operators to enter programs and data slowly, by hand, on paper cards or paper tape. They would put the cards into a "reader," then wait 20 minutes or so for a result. No one imagined that computers could work instantaneously. And no one, it seemed, could see the benefits of devoting a computer's power to tasks such as writing and outlining.

Today, the spiritual descendants of Engelbart's NLS, with its mouse and graphic display and "writing machine," sit on desks around the world. And with the arrival of the Internet and local-area networks and "groupware" like Lotus Notes, it seems that much of Engelbart's vision has become reality. Yet the core ideas of his work on "augmentation" remain unrealized. The personal computer has allowed us to work better, but we still work, for the most part, alone. "Relative to what our potential is, we can go as high as Mt. Everest, and we're only at 2,000 feet," he says.

So Doug Engelbart continues to work toward achieving those goals he articulated to himself 50 years ago. Humans still need new tools, and new methods of working, to cope with the ever-increasing complexity and urgency of the world. He now leads an organization called the Bootstrap Institute, which is dedicated to raising the awareness of the challenges facing humans, as well as the potential for new systems to meet those challenges. "Bootstrapping is sort of like compound interest," he says. With the proper approach, an organization's progress in making itself work better will build exponentially, as each advance improves on earlier improvements. Today's computers and the Internet and the Web are small advances, but the problem is still huge. "If we humans don't learn how to be collectively smarter as fast as we can," Engelbart explains, "there's going to be a better and better chance that the human race is just going to crash. That's what keeps us going."

itself, a super-microfilm reader. But what was revolutionary about the Memex was not that it stored information, but the way in which one accessed it. The essential feature of the Memex, Bush wrote, was its associative indexing, "whereby any item may be caused at will to select immediately and automatically another." Bits of information would be intertwined, becoming "a trail of [a user's] interest through the maze of materials available to him." If ideas and images and documents could be connected in this way, it might help manage the flow of that great quantity of information that threatened to smother science and business.

The "trails" and connections of the Memex are close cousins to what we now call hypertext. And just as today's Internet visionaries predict new forms of content blossoming from such connections, so too did Bush. "Wholly new forms of encyclopedias will appear," he wrote, "ready made with a mesh of associative trails running through them. . . . The lawyer has at his touch the associated opinions and decisions of his whole experience, and of the experience of friends and authorities. The patent attorney has on call the millions of issued patents, with familiar trails to every point of his client's interest. The physician, puzzled by a patient's reactions, strikes the trail established in studying an earlier similar case, and runs rapidly through analogous case histories, with side references to the classics for the pertinent anatomy and histology." Some 50 years before its time, Bush was seeing, however faintly, the promise of the World Wide Web.

AL GROSS

One day in the late spring of 2000, two climbers set out to scale Mount Hood, the highest peak in Oregon. The ascent did not go well—a few hours into it, a rock slide injured the two men, seriously enough that one of them couldn't move. Fortunately, they were carrying a small communication device, and they began sending out pleas for help. Seventy miles away, a seven-year-old boy was holding a similar device, and he heard a scratchy voice from the mountain. The boy and his father assured the two men that rescuers were on their way, then helped guide the Air Force Reserve's Rescue Wing to the injured climbers.

The climbers' personal communication device wasn't a satellite phone, or a cell phone, or anything particularly new. It was a walkie-talkie, little changed since its invention more than 70 years earlier.

The story of the walkie-talkie, and its inventor, Al Gross, begins with a boy almost as young as the Oregon rescuer. In 1927, when he was just nine, Gross and his family were crossing one of the Great Lakes on a steamship. The inquisitive Gross was exploring the ship and happened upon the radio operator's cabin. He was instantly enthralled. "I saw all that electrical stuff the guy had there," Gross remembered. "He put me on his lap and he put the earphones on my ears and I heard the dots and dashes and that intrigued me all over. Boy! I had to learn what he was doing and how he did it."

After some persuading, Gross's father bought him a crystal radio set, which let him listen to the airwaves. His next step was to become an amateur radio operator, so he could talk to other radio buffs miles away. By the age of 15, Gross was building and experimenting with his own radios. He also began studying the work of the radio pioneers, including Nikola Tesla, Heinrich Hertz, and Guglielmo Marconi.

When Gross walked into that ship's radio cabin, it had been barely 25 years since Marconi had sent the first wireless message across the Atlantic Ocean, an event that ushered in the modern era of telecommunications. (And it wasn't until radio distress calls had helped save some 700 people from the sinking *Titanic* in 1912 that people really started paying attention to the new

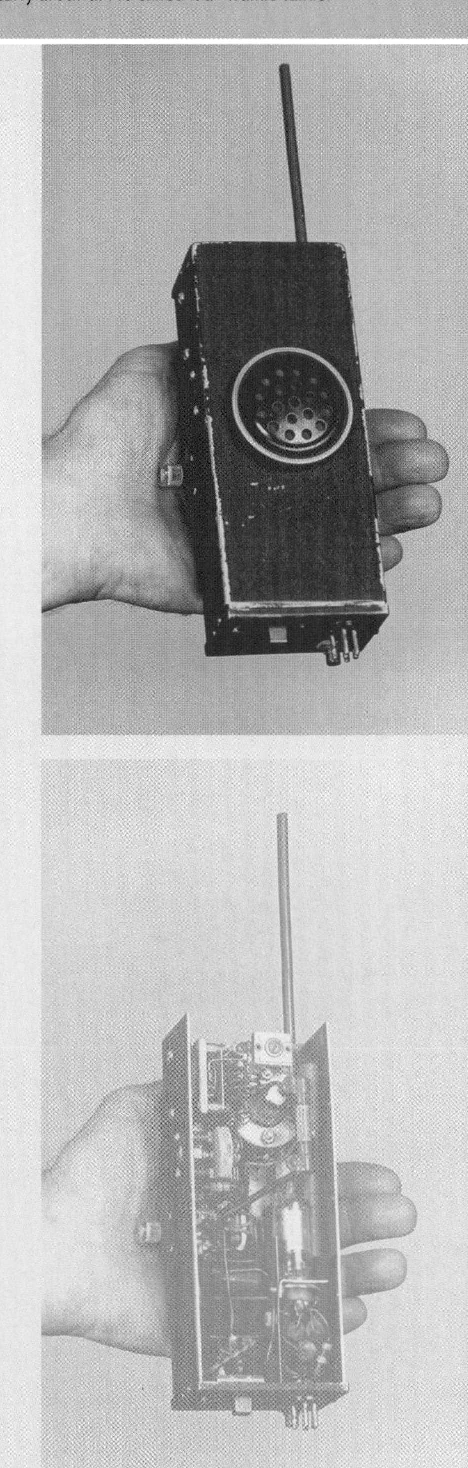

Al Gross fell in love with radios when he was just nine years old; a decade later, he built a two-way radio that could he could carry around. He called it a "walkie-talkie."

medium.) So when he began to tinker with his equipment and think of ways to make it better, there was a lot of fresh exploring to do.

By the mid-1930s, the young radio operator had in mind one particular improvement he wanted to make. "I didn't like to sit in the house and talk," he explained. "I wanted to be mobile." The solution was obvious: Build a radio transmitter and receiver (a transceiver) that was small and light enough to carry around. Gross began by searching out smaller and smaller components. He also started thinking about what radio frequency to use. At the time, broadcasters weren't using frequencies higher than 100 megahertz—about the frequency used for FM radio today. Experimenting at home, Gross was able to make miniature vacuum tubes operate at up to 300 megahertz, far higher than anything anyone else was using. (Radio signals now fill a range from about 535 kilohertz, used for AM radio, up to 2 gigahertz, used in newer cordless phones.) By 1938 he had perfected a portable radio that he could walk around with while he was talking—a walkie-talkie, his radio friends called it. The new device was just eight inches long and a little more than two inches wide.

About this time Gross began his college education, studying electrical engineering at the Case School of Applied Sciences (now Case Western University) in Cleveland. But World War II was on the horizon, and the Office of Strategic Services (or OSS, the forerunner of the CIA) was keenly interested in new kinds of communications devices. A radio operator who had heard about Gross's walkie-talkie brought it up at the OSS, and Gross traveled to Washington, D.C., to demonstrate it. Very soon he was working on the war effort, developing a top-secret communications system called Joan/Eleanor, named after his daughter and wife.

The system had two elements: Joan was a portable device much like his walkie-talkie, and Eleanor was a larger and heavier receiver that was installed in planes that could fly at high altitudes. OSS agents working behind enemy lines would carry a Joan unit, which weighed just four pounds and worked on batteries. At certain times, they would broadcast straight up into the air, where

By the time he was 18 years old, Gross had built his own short-wave radio station, which he installed at his home.

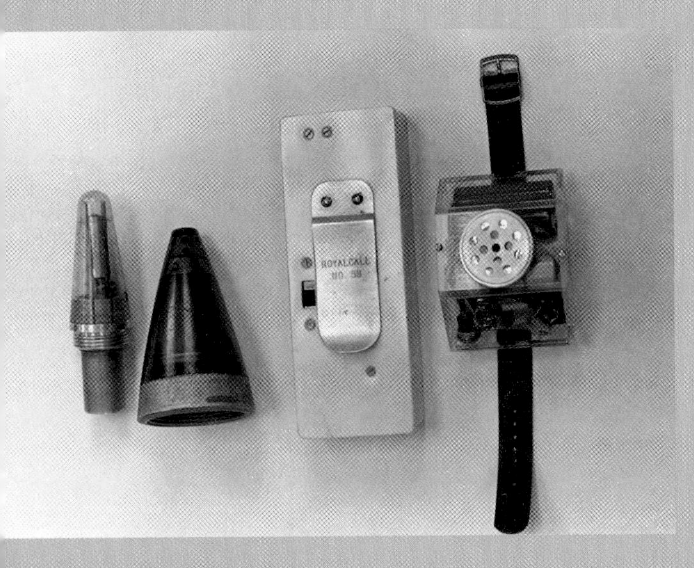

Three of Gross's inventions: from left to right, a proximity fuse for use in weapons; the world's first pager; and a two-way wristwatch radio that inspired the one worn by Dick Tracy.

The walkie-talkie itself went though several incarnations, each getting smaller and lighter.

an unseen plane would pick up their messages and send news back down to the ground.

The Joan/Eleanor system had huge advantages over other ways for agents to communicate with headquarters. First, it was light, small, and concealable. But just as important, because it operated on the unexplored high frequencies, the enemy—in this case, Germany—couldn't listen in. And even if they did try, they'd have to be flying directly overhead, since the Joan unit was designed to transmit straight up in a very narrow band. And the portable units put out a strong signal, so the listening planes could fly as high as 30,000 feet, beyond the range of most German countermeasures. The system was put into use in Europe in late 1944, and was used successfully until the end of war. The U.S. Joint Chiefs of Staff called it one of the "most successful wireless intelligence gathering operations, saving millions of lives by shortening the war."

After the war, Gross wanted to extend his new device beyond the world of secret agents. The Federal Communications Commission (FCC) soon allocated a range of frequencies for walkie-talkie use—the Citizens Radio Frequency Band, which would eventually lead to the CB Radio craze of the 1970s. Gross began manufacturing walkie-talkies, and he sold more than 100,000 of them. The challenges and pressures of such a large operation were not what interested him, though, and he licensed the technology to other companies.

Gross was continually trying to improve on his walkie-talkies, and in 1948 he built a radio that could fit on a man's wrist. Chester Gould, who drew the daily comic strip "Dick Tracy," visited Gross's workshop and was impressed with the prototype. That October, Dick Tracy began wearing his now-famous two-way wristwatch radio.

Next, Gross created an even smaller communication device, which would become known as the pager. Essentially a miniaturized two-way radio, the little silver boxes could be set to respond to specific signals—a short message would be broadcast to every pager, but only a specific person's device would recognize it and beep. Gross had high hopes that doctors and nurses would adopt his new invention as a way to keep in touch with their hospitals or offices. But the pager met with resistance; nurses feared it might disturb patients, and one doctor, Gross recalled, complained that it would interrupt his golf game.

By 2000, Al Gross saw the world flourishing with wireless communications devices. Cell phones and pagers are constant presences in our lives, and anyone can gain access to the World Wide Web from a personal digital assistant. Although Gross was not directly responsible for cellular technology, the current wireless revolution is in many ways the culmination of his vision for instant, portable communication. In July 1945, just after World War II ended in Europe, E. K. Jett, the commissioner of the FCC, wrote a story for the *Saturday Evening Post* that described a world using a version of Gross's walkie-talkie.

The "handie-talkie," Jett explained, could be carried around by anyone, even used in cars. Its use would range from deliverymen receiving orders to families staying in touch to emergency calls for help. "One can picture a young woman motorist riding alone at night on a lonely road," Jett wrote. "A car comes roaring down an intersection, sideswipes her coupe, and crashes against an abutment. . . . She turns to the handie-talkie, which is slung camera-like over her shoulder. . . . She pulls out the antenna, spins the dial to Citizens' Radio distress frequency . . . and in a few minutes an ambulance arrives."

In 1977, Gross (center) appeared on the TV quiz show "To Tell the Truth."

ERNA SCHNEIDER HOOVER

Before the 1950s, the process of making a telephone call often involved simple personal contact. In many places, when you picked up the telephone, an operator would ask, "Number please?" After you responded, the operator would make a physical connection with a wire to the switchboard and your call would go through. It was a comforting, human aspect to a relatively new technology, and it worked pretty well, at least for a while.

As the U.S. economy boomed in the decade following World War II, people and businesses began to use the telephone more and more. Technological advances increased the load on the system: Beginning in 1951, microwave relay stations increased long distance service, and in 1956 the first transatlantic telephone cables started carrying international calls. Bell Telephone's own projections made it clear that the continually increasing demand for telephone service would one day—and probably soon—overwhelm the system. The answer to this looming problem was the same as in many other industries at the time: computerized automation. Around 1954, Bell Labs, the research arm of the Bell Telephone company, began working on automating the process of connecting calls, known as switching.

More than 10 years and $500 million later, Bell Telephone unveiled the first large-scale electronic switching system, or ESS. Although the company advertised that "more than 2,000 man-years of research" had gone into ESS, one of the key figures in its development was a woman, Dr. Erna Schneider Hoover. A medievalist, logistician, working mother, and computer programmer, Hoover designed the basic architecture of the switching system, which could automatically route hundreds (and later, thousands) of calls in a few seconds. This work earned her one of the first patents ever awarded for software and a job as a research supervisor at Bell Labs, the first time a woman had held such a position.

Erna Schneider grew up in Irvington, New Jersey, where she was born in 1926, and moved to Massachusetts in 1944 to attend Wellesley College. She studied medieval history and philosophy there, earning her bachelor's degree in 1948. Since Hoover intended to both work and raise a family, she focused

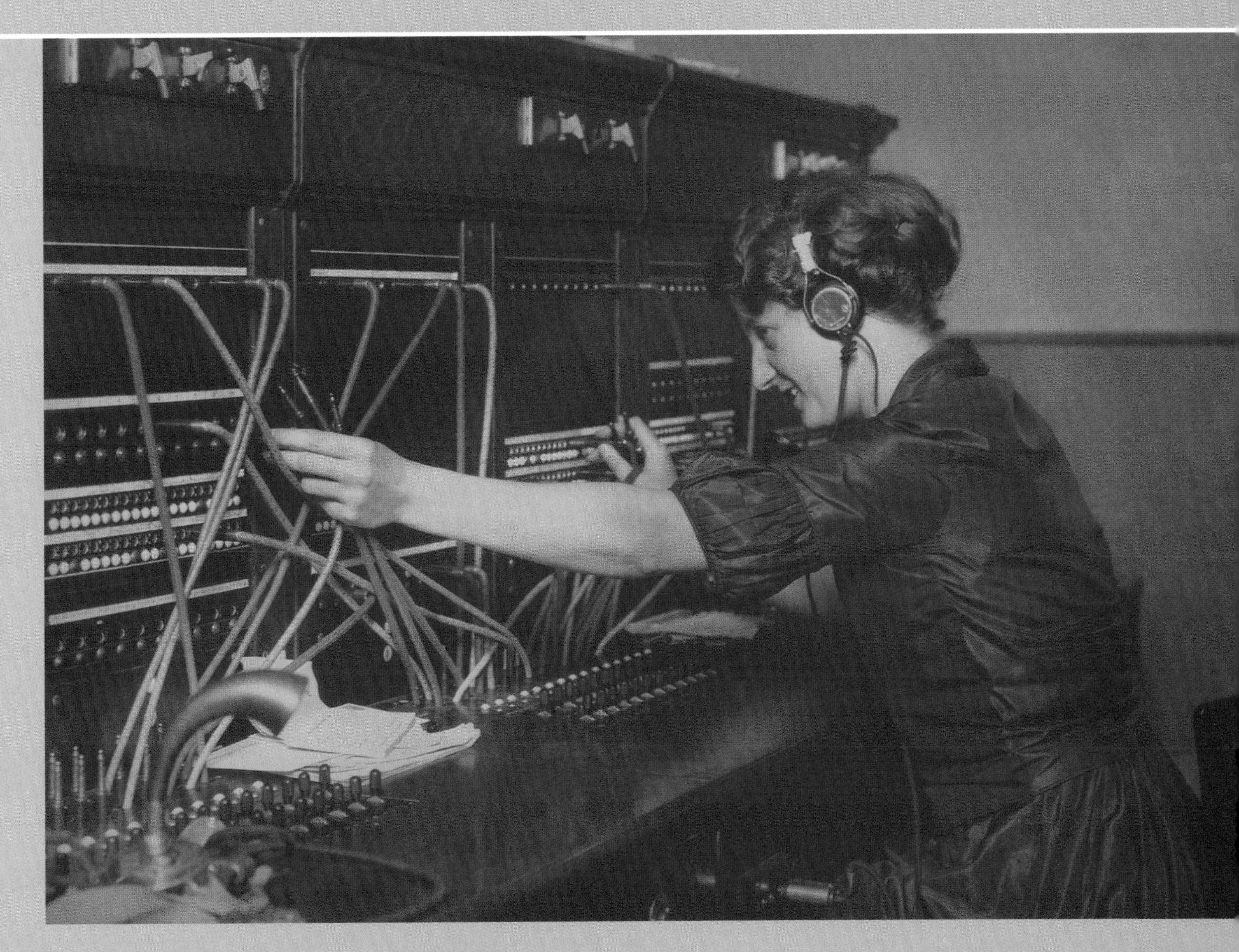

The process of "switching" (connecting) calls before automation depended on operators working at switchboards. In the 1950s, Bell Telephone introduced direct dial, and in the 1960s it added completely automated service.

BELL SYSTEM
MANUFACTURED BY
Western Electric
NUMBER ONE
ELECTRONIC SWITCHING SYSTEM

on getting a position in academia, where the hours were more flexible. In 1951, she earned her Ph.D. in logic and the philosophy of science at Yale University.

As a graduate student Hoover encountered the challenges of sexism. When she earned her doctorate at Yale, she was told that she would have been kept on to teach there "if I had been a man." However, her experience and success at college helped her deal with this prevailing attitude. "The academic grounding Wellesley gave me left me feeling that I could learn almost anything," she told the *Wellesley* in 1990. "I had a supreme sense of self-confidence from professors who were role models and students who showed they had ability to run things themselves. I found [college] liberating after high school, where girls were supposed to act dumb even if they were not."

After Yale, Hoover taught philosophy for three years at Swarthmore College in Pennsylvania. She married Charles Wilson Hoover, Jr., an engineer who had just been hired by Bell Labs, and in 1954 she, too, joined the research team at Bell. When Hoover got there, she remembered, "they were planning to develop a telephone switch that could be controlled by electronic computer." The idea of computer control was the key: Problems could be fixed and adjustments made from a central computer, rather than by altering or swapping the electromechanical components then in use. A computer could also check its own performance, testing and fixing itself.

Computer science was still in its infancy at the time, and Hoover's seemingly arcane academic specialty, symbolic logic, was suddenly in high demand. A computer is essentially made up of many logic gates; vacuum tubes were the first such gates used, although by the time Hoover got involved, Bell Labs' transistor, invented in 1947, was replacing tubes. (The integrated circuit would replace the transistor during the following decade.)

The first electronic switching system (ESS) was installed in Succasunna, New Jersey, in 1965. On the table are the ESS's major components; the square object second at left is one of its mechanical memory cards, made before silicon memory chips.

Logic gates govern how information bits, or at their most basic level, electrons, move through circuits. These gates obey strict rules of logic; the three essential gates are AND, OR, and NOT. In an AND gate, if the two arriving bits are 1s, the gate produces a 1. With an OR gate, if either (or both) of the arriving bits is a 1, it produces a 1. And with a NOT gate, a single input is reversed; a 1 produces a 0, and vice versa. There are four other gates that are variations on the basic three that combine their logic; for example, the NAND gate marries NOT and AND, producing 1s where an AND would produce 0s.

“I found [college] liberating after high school, where girls were supposed to act dumb even if they were not.”

By creating sequences—circuits—of these gates, an engineer (or a logistician) can manipulate data in useful ways. A particular combination of 12 gates, for instance, makes a circuit that can add two bits; a series of eight of these circuits can add a whole byte. Today's processors, with more than 10 million gates, can quickly manipulate large quantities of data.

Since an electronic phone switching system would need to make hundreds or thousands of logical decisions every minute, Hoover was well suited to the task. (Computer programs take advantage of the same strict logic, although high-level programming languages allow programmers to use much more complex arguments.) In her description of the ESS, she noted that most of the computer's time would be spent “performing logical operations. . . . [An] arithmetic unit is not needed, but a wealth of logic instructions are.”

Hoover conceived of the basic logical architecture of the Electronic Switching System while in a hospital bed, following the birth of her second child. The ESS had two main components: a large memory bank (which relied on physical, magnetic storage, since it predated memory chips) and a “central control,” which contained the computer and its program.

One of the main challenges in switching calls was that they did not come in at regular intervals; at night, for example, it was simple to take care of the few connections to be made. During peak hours, though, a system could easily be overwhelmed. "Unlike a typical computation center," Hoover wrote, "in which work is scheduled in job shop fashion, the ESS cannot control the time when it must process calls, because telephone customers expect to receive prompt service at any time."

Hoover created a set of logical rules for how the system dealt with calls, and another for how the computer divided its attention among the calls coming in. The system she and her colleagues built used some timing tricks to deal with more calls than it otherwise might, by queuing extra calls and switching its attention from call to call, since the computer worked much faster than callers could. "Once having determined what sort of signal has come in," she noted, "the system can afford to take more time before making the chain of logical decisions which determines how to treat the call." Because all of those logical decisions were made by the computer's program, rather than the phone company's physical equipment, changes in phone service could be made easily and quickly.

The finished ESS system, which debuted in Succasunna, New Jersey, in 1965, was one of the first uses of large-scale computer automation to reach the public, although its function was so transparent that most callers never realized it. (Computer-controlled switching did, however, allow several notable new telephone features such as call forwarding and call waiting.) While that first installation covered only 200 phone subscribers, within a decade 6 million phones were being switched electronically. Today, virtually all phone calls are routed through computerized switching centers, although many have been upgraded to the next generation of "digital switching."

Hoover continued her research work at Bell Labs after the ESS was completed. During the 1960s, she contributed to the antiballistic missile defense system, which successfully intercepted mock ICBM warheads before the program was banned by treaty. In 1978, she became the first female head of a research division at Bell Labs, where she worked on the application of artificial intelligence to communications systems.

In 1971, Hoover was granted a patent for the software governing electronic call switching. This was not only one of the first patents ever granted for a computer program, it was also proof of one of the larger meanings behind Hoover's invention. Although it involved a complicated computer system, the key achievement of the ESS was its software; as such it represented one of the first triumphs of software over hardware. While today software is almost always more useful, impressive, and important than the hardware it runs on, in the 1960s Hoover's work was a sharp (if not immediately noticed) contradiction of the belief that bigger computers were the answer to any information-processing problem.

GRACE MURRAY HOPPER

Grace Murray Hopper discovered the pleasure of finding out how things work when she was just seven years old. Intrigued by one of her family's alarm clocks, she took it apart, spilling out its springs and gears. Still curious, she opened up another clock, and another, until she had disassembled seven of them. Although it's not clear whether she was able to put any of them back together, the experience began a lifelong quest for knowledge, a quest that made her one of the great pioneers of computing. Working with some of the world's first computers, Hopper established basic methods for writing software and created the first high-level programming languages. Her ideas were applied to the development of all subsequent languages, including those used today.

Grace Murray was born in 1906 in New York City. Her father was an insurance broker with a love for books, and her mother was a housewife with a deep interest in mathematics. While many parents would be dismayed by the destruction of their alarm clocks, Hopper's encouraged her technical interests. As a young girl, she would sometimes accompany her grandfather, a civil engineer, as he surveyed the city to plan new streets. And while most young women were not then encouraged in academic pursuits, Hopper's father thought that she should get the same education as a boy. In 1924, she entered Vassar College, one of the country's top schools for women. There Hopper studied mathematics and physics and graduated in 1928 with honors; two years later she earned her master's degree in mathematics at Yale University. The same year, she married Vincent Hopper, an English professor who joined her at Vassar.

Hopper became a well-liked professor of mathematics at Vassar; at the same time she was studying for a Ph.D. in that discipline, which she received from Yale in 1934. She continued teaching for the next eight years, and she might have remained teaching at Vassar much longer but for the start of World War II. Hopper had always been deeply patriotic—her great-grandfather had been an admiral in the navy, and during World War I she had knitted caps and socks for American servicemen. In 1942, after her husband and brother had joined the air force, Hopper decided to join the navy. It was not

“Humans are allergic to change,” she declared. “They love to say, ‘We’ve always done it this way.’ I try to fight that.”

Hopper devoted much of her life to the navy. She served in the women’s reserves from 1943 to 1946, and then as an officer from 1967 through 1986.

9/9

0800 Arctan started
1000 " stopped - arctan ✓ {1.2700 9.037 847 025 / 9.037 846 995 correct
13″ UC (032) MP - MC ~~2.130476415~~ 4.615925059(-2)
(033) PRO 2 2.130476415
correct 2.130676415
Relays 6-2 in 033 failed special speed test
In relay " 10,000 test.
Relays changed
1100 Started Cosine Tape (Sine check)
1525 Started Mult + Adder Test.
1545 Relay #70 Panel F (moth) in relay.
First actual case of bug being found.
1630 Arctangent started.
1700 closed down.

Relay 2145
Relay 3376

The first computers were made before transistors and depended on relays and other mechanical devices. When the Mark II computer Grace Hopper was using during the summer of 1945 failed, she tracked down the cause: a moth caught in a relay, the first computer "bug."

easy for her to get in. In addition to being too old at 34, she was underweight, and her job as a professor was considered essential. But with special permission, Hopper joined the navy's women's reserves in 1943.

After basic training, Hopper was assigned to work on one of the world's first computers, the navy's Mark I, at Harvard University. The Mark I was a large and primitive machine. Some 51 feet long and eight feet high, the computer relied on electromechanical switches—relays—to process numbers. (The transistor wouldn't be invented for a few more years.) Hopper was immediately excited to work with the Mark I. "It was the fanciest gadget I'd ever seen," she later said. "I had to find out how it worked."

Hopper's job was to program the Mark I to calculate the trajectories of missiles. But the Mark I, like all early computers, was not easy to use. First, Hopper had to figure out the equations that described how a missile would fly. Then she had to break those down into basic steps simple enough for the computer to understand. Next, Hopper would write these steps down as instructions for the computer, in a language it could understand, using just ones and zeroes. Finally, those instructions would be encoded onto a long piece of paper tape—a punched hole for "one," no hole for "zero." It was a time-consuming and error-prone process, but it worked most of the time, and Hopper, her colleagues, and their Mark I (and later the Mark II and Mark III) were able to produce missile trajectories much faster than human calculators could.

In the summer of 1945, Hopper was working with the Mark II when the machine stopped functioning. Searching for a mechanical failure, she discovered a moth squashed in a switch. Hopper kept a log book of the computer's operation and taped the moth into it, noting that it was the "first actual case of a bug being found." From then on, whenever she and her colleagues were trying to fix the computer or a program, they called it debugging.

In 1946, Hopper was forced to retire from the navy. She remained at Harvard for a few years, then joined the Eckert-Mauchley Computer Corporation, which was building the first commercial computer, the UNIVAC. This machine was smaller, faster, and more flexible than the Mark I, and with it Hopper began to explore new ways to program computers. Until then, every part of a program had to be re-entered every time it was used, even if the computer had done that operation—dividing two numbers, for example—many times before. The programmer and the computer had to start from scratch each time.

Hopper thought this was a waste of time and resources. She began to collect pieces of programs that could be used again and again. "I kept them in a notebook," she recalled. "But to put them in a new program I still had to copy them [by hand]. So I decided to make a library of all these pieces of code and I'd give them each a name and then I'd tell the computer to put them together, copy them, and add them to the [computer's memory]." What Hopper had created was the first compiler—a program that translated these "libraries" into binary code for the computer. With it, a computer could "automatically" write its own programs, based on the instructions fed into Hopper's compiler.

At first, the "names" of the pieces of code were mathematical functions, since at that point most computers were used for scientific or military purposes. But Hopper realized that there were far wider applications for the machines, particularly in businesses that had to process large amounts of information. "So I decided data processors ought to be able to write their programs in English, and the computers would translate them into machine code," she explained. "I could say, 'Subtract income tax from pay' instead of trying to write that in octal code or using all kinds of symbols."

None of her colleagues believed that computers could handle programs written in English. That kind of attitude always bothered Hopper. "Humans are allergic to change," she declared. "They love to say, 'We've always done it this way.' I try to fight that." One way she fought was by hanging up a clock that ran counterclockwise; it still told time, just in a different way.

After two years of work, Hopper did make a compiler that let the UNIVAC understand English commands such as Compare, Transfer, Price, and Write. Called Flow-Matic, it was one of the very first high-level computer languages, and was soon put to use by insurance companies, the military, and other organizations.

By the late 1950s, several companies were beginning to make computers. However, Flow-Matic worked only on the UNIVAC. Hopper and others in the computer industry recognized that they needed a programming language that could be used on any computer. In 1959, Hopper became part of a committee that was planning to make just such a language, called COBOL (for COmmon Business-Oriented Language). Although she did not work on COBOL herself, her Flow-Matic served as one of its main models, and her idea about using common English commands was one of its programmers' guiding principles. Introduced in 1960, COBOL became the most common and most popular computer language, the basis for most programming as the United States and the world became more and more computerized. Today, many important large-scale programs written in COBOL are still in use; during the Y2K scare in 1999, companies had to scramble to find programmers who could work on their long-running COBOL software.

In addition to being a pioneer programmer, Hopper was a highly effective leader. One of her passions through the late 1950s and 1960s was persuading businessmen and her colleagues in the computer industry to use English-based programming languages. In 1967 she was recalled to navy service because no one else could coordinate the service's different COBOL programs; she did it by creating a standard version of the language, USA Standard COBOL. Although she was supposed to serve for only six months this time, she remained an officer until 1986; by then she had been promoted to Rear Admiral, and was the oldest active navy officer. Grace Hopper died in January 1992 at the age of 85.

One of Hopper's favorite things to teach was the importance of taking chances, of ignoring the usual way things are done. "A ship in port is safe," she liked to say to colleagues or students. "But that's not what ships are for. Be good ships. Sail out to sea, and do new things."

RAYMOND KURZWEIL

RAYMOND RAYMOND KURZWEIL

Raymond Kurzweil has been making computers smarter since he was in high school. As a teenager, he built a computer that could analyze melodies and then create new ones based on that information. In 1976, he introduced the first commercial product based on artificial intelligence, a scanner-character-recognizer speech synthesizer that could read books aloud to the blind. A decade later, he created the first large-vocabulary speech recognition system. As an inventor on the cutting edge of technology, Kurzweil has also become a student of technological change, using that knowledge to speculate on and predict the future of computing and technology.

Kurzweil grew up in Queens, New York, where he was born in 1948. Although his parents were artists—his father a musician and conductor and his mother a visual artist—academics ran in the family. "There were a lot of doctors and scientists and Ph.D.s," he notes. "Including the women. My mother's mother was one of the first female Ph.D.s in chemistry." And he had at least one inventive ancestor, his father's father, who built and marketed automation systems.

His inventive impulse was clear at the age of five, when he began building his own model boats, cars, and (failed) rocket ships. The young Kurzweil's goal was to be a scientist—an important scientist—and to him that meant being an inventor. It soon came to mean inventing things with computers. "I would go down to Canal Street in New York and buy surplus electronics," he remembers. He built a simple computing device when he was 12, and he also learned how to program. Kurzweil's uncle, an engineer at Bell Labs, helped inspire these pursuits, but Kurzweil did most of his learning and experimentation on his own.

When Kurzweil was 15, he began his first project involving pattern recognition—teaching machines how to see and understand patterns in information. The result was a computer that could write original melodies based on patterns it saw in the work of a composer. It also landed him on the game show "I've Got a Secret," where he played a machine-written composition on

In 1976, Ray Kurzweil's company introduced the Reading Machine, a combination scanner, character recognition system, and speech synthesizer that could read books aloud.

piano. (The panelists had to guess his secret, that a computer had written the piece.)

While in high school, Kurzweil began corresponding with Marvin Minsky, the pioneer of artificial intelligence at the Massachusetts Institute of Technology. This friendship helped convince him to attend MIT, where he began studying computer science. But at the time MIT had only eight or nine computer classes; Kurzweil took them all during his first two years, then added a second major in creative writing. (He continues to combine technical and literary pursuits.) The classes Kurzweil took occupied only part of his time at MIT; he was also running a company, which used a computer to match high school students with colleges. He eventually sold his company to Harcourt Brace and World for $100,000 plus royalties.

In 1974 Kurzweil founded his first major company, Kurzweil Computer Products, to tackle a long-standing challenge: creating a machine that could read printed text in any font. This was a classical problem of pattern recognition; other inventors had made reading machines, but to Kurzweil they "basically just matched pixels against a perfect template."

Kurzweil and his colleagues wanted to rely on the patterns of letters rather than a template. "We wanted to really attack the general problem of recognizing characters, to consider a character to be an image with a more abstract definition," he says. "What makes a capital A inherently a capital A?" As he explored how to achieve this, he developed some basic principles that he has followed in all subsequent pattern-recognition problems: use multiple approaches; learn from the real world; and make a system that can itself figure out patterns.

He defined several different ways of recognizing a letter; some used mathematical transformations of the image, while others analyzed invariant features—the constant properties of a character. A capital A, for example, has a triangular looped portion; line segments that run from the top to the bottom corners; and a cross bar. Then Kurzweil wrote an "expert" program that weighed the strengths and weaknesses of the different recognizers and decided from that information what the character

RAY KURZWEIL'S CYBERNETIC POET

Ray Kurzweil has applied pattern recognition to literature as well as language. By reading and analyzing a selection of an author's writing, his program "Ray Kurzweil's Cybernetic Poet" can create a "language model" of how he or she writes. Using these models, the program can write poems that replicate the rhythm, style, and structure of the original poet. It can also combine the characteristics of several poets in one work.

Here are a few examples of the program's cybernetic poetry.

A Wounded Deer Leaps Highest

(after reading poems by Wendy Dennis and Emily Dickinson)

A wounded deer leaps highest,
I've heard the daffodil
I've heard the flag to-day
I've heard the hunter tell;
'Tis but the ecstasy of death,
And then the brake is almost done,
And sunrise grows so near
sunrise grows so near
That we can touch the despair and
frenzied hope of all the ages.

The Saxophone Player

(after reading poems by Ray Kurzweil and Pat Sherman)

The saxophone player
lives alone,
blows
lives alone,
blows
a swinging door
splendid silence
prophetic poses
splendid silence
prophetic poses
of a prayer and the walls.

was. "Any of [the individual recognizers] could do the whole job," he notes, "but not as accurately as the whole system." Kurzweil's recognition system was also able to learn from the different characters it saw. "You can only get so far if you just define high-level properties and say an A has these particular features," he says. "You'll find real world images that don't have all those properties. So you need a system that can actually learn what real images are like."

Just as he was getting his system to work, Kurzweil met a blind man on a plane. As they talked, Kurzweil remembers, "he was explaining, 'Blind people aren't handicapped. I travel around the world and I participate in meetings and I talk on the phone and there's really no limitations.' Except he did have a problem accessing ordinary printed materials. If he could access ordinary print, that would really address the primary limitation he faced." The character recognition project suddenly had a much larger goal: to build a machine that would read to the blind.

Although Kurzweil's recognition system was doing a good job reading characters, two problems remained in building a system that could scan a book and speak it out loud: Neither the flat-bed scanner nor text-to-speech synthesizers had been invented yet. Within about a year, though, he and his engineers accomplished both of these tasks, using equipment that now seems very primitive. The scanner, for instance, could only see a small part of the page at a time, so it had to move back and forth and then down the text to read it all. On January 13, 1976, the Kurzweil Reading Machine was demonstrated to the press. That night, CBS news anchor Walter Cronkite did not deliver his usual sign-off, "And that's the way it is." Instead, the Reading Machine spoke it.

The musician Stevie Wonder, who is blind, heard the announcement that day and soon bought the first production model of the Reading Machine. That purchase began a long friendship between Wonder and the inventor, which, in the early 1980s, spawned Kurzweil Music Systems, a line of music syn-

Moon Child

(after reading poems by Kathleen Frances Wheeler)

Crazy moon child
Hide from your coffin
To spite your doom.

Tomcat

(a haiku, after reading poems by Randi and Kathryn Lynn)

An old yellow tomcat
lies sleeping content,
he rumbles a heart

Dark

(after reading poems by Kathryn Lynn)

I am the dark.
on
the night
The
past of love
gone stale.

Musician Stevie Wonder bought the first Reading Machine and later worked with Kurzweil to develop a music synthesizer that accurately re-created the sounds of acoustic instruments.

thesizers that mimicked traditional instruments so well that most listeners couldn't tell the difference.

In the 1980s, Kurzweil also began working on another classic problem of pattern recognition, the opposite task of the Reading Machine's: turning speech into text. For years, a system that could manage this was thought of as a Holy Grail of artificial intelligence. "All of our intelligence is embodied in our language," Kurzweil says. "Speech embodies not only textual language but also the intelligence we put behind the inflection and the emotion we can represent and the cadence of our voice."

Kurzweil used the same principles as in character recognition. The advantage of starting ten years later was that computer speed had increased exponentially—which was essential, because speech recognition takes enormous processing power. Different ways of recognizing a sound were employed, along with an expert program to decide which sound it really was. The system Kurzweil designed was also able to develop its own models of the patterns found in speech. Rather than telling the system that there were, say, exactly 44 phonemes in the English language, he says, "we let it figure out itself how many different patterns it finds and what those patterns look like. It will actually develop different models for different dialects."

The speech-recognition system was first introduced in 1987, and its early users were people with disabilities that prevented them from typing. Doctors also began using it, since the limited vocabulary used in medical records lent itself to recognition. The technology has come into its own in the past few years, though, with about a million systems now installed. Soon, Kurzweil says, speech will become an important interface between humans and machines.

Making such predictions is another of Kurzweil's talents; many appear in his two best-selling books, *The Age of Intelligent Machines* (1990) and *The Age of Spiritual Machines* (1999). "Writing is an opportunity to invent with the resources of the future," he says. "So I can invent using the computers and communications resources of 2010 or 2020." To invent the future accurately, though, he had to have a very good idea of what technology would be available. One key was recognizing that the rate of technological change is accelerating, doubling every 10 years. So what might seem to be a 100 years in the future is really only 25 years away.

Still, this attention to how fast things change has as much to do with Kurzweil's current projects as it does to those of 2025. "I became a student of technology trends as part of the discipline of inventing," he says. "You want your project to make sense when it's finished, not when it started."

When he was only 15 years old, Ray Kurzweil appeared on the game show, "I've Got a Secret." Kurzweil's secret: the music he played on the piano had been written by a computer, which he had programmed to create new melodies from existing works by human composers.

CARVER MEAD

"I get interested in particular topics," explains physicist, engineer, and innovator Carver Mead, "because the way people are going at it doesn't make any sense to me. That drives me nuts, and that's what gets me motivated to figure out a better way to do it."

The "better ways" Mead has figured out have had far-reaching impacts. Some of the topics he has become interested in include the physics of semiconductors, microchip design, biological computing, and digital imaging. And in each of these realms, Mead has made tremendous advances and discoveries—developing, for instance, a method for designing chips that is now used for almost all high-performance microprocessors. One of his most recent interests, digital imaging, resulted in the recent unveiling of a digital camera far more advanced, and more like photography on film, than any previous digital camera.

Mead's childhood revolved around the technologies that would eventually become his career: electricity and electronics. Born in Bakersfield, California, in 1934, Mead grew up in Big Creek, a "camp" in the Sierra Nevada where his father ran a hydroelectric plant. His first fascination was with the power plant, where, he remembers, there was a "palpable energy." When he was about 13, though, an amateur radio operator moved into the camp and Mead was soon hooked on electronics. "There was something special about radio," he says. "That you could communicate over large distances, the whole propagation of radio waves, that was just fascinating to me."

World War II had recently ended, and the market was flooded with inexpensive surplus electronics. Mead bought as much of it as he could and spent his spare time tinkering. Between the plentiful equipment and the quick advances being made, he says, "It was just a fabulous time to be getting into electronics." In 1952 Mead began studying electrical engineering and physics at the California Institute of Technology. It was also a fabulous time to begin studying at CalTech. Physicist Richard Feynman had just joined the faculty; Mead took his mathematical physics class and also began a lifelong friendship and collaboration with him. His freshman chemistry professor was Linus

Pauling. "That [class] was my start in quantum mechanics," Mead remembers. "He just knew that stuff absolutely cold."

Mead earned all of his degrees at CalTech—a bachelor's in 1956, a master's in 1957, and a Ph.D. in 1960—and was subsequently hired as an instructor. (He is currently the school's Gordon and Betty Moore Professor of Engineering and Applied Science.) Mead's initial research project as a faculty member led to his first major invention, what is now known as the HEMT transistor. "I studied the detailed physics of the contacts between metals and semiconductors," he explains. "It doesn't sound very exciting, but it actually is." His investigation into the physics of semiconductors convinced him that he could make an entirely new kind of transistor that could work better. Mead was correct; his transistor was able to work at much higher frequencies than previous ones, and they became a key element in modern microwave communications, including satellites and cell phones.

As his research into metal-semiconductor junctions continued, Mead began to look into a phenomenon called electron tunneling. When insulating materials get thin enough, he explains, "the fact that the electron is actually a wave starts to matter. Electrons can penetrate through the very thin layers of these materials." Because circuits are based on the orderly and predictable flow of electrons, this behavior could be troublesome. One day, Mead was talking about this phenomenon with Gordon Moore, who was then at Fairchild Semiconductor. "He said, 'Carver, doesn't that mean that if you make a device sufficiently small, that it won't really work anymore?'" Mead remembers. "I said, 'Yeah, I think that's what it means.' And he said, 'Why don't you work on how small that is?'"

In the late 1960s, after he had figured out how small integrated circuit components could be made, Carver Mead realized a new system was necessary to design chips as they became more complex. His solution was Very Large Scale Integration (VLSI), a simpler and more systematic approach to chip design.

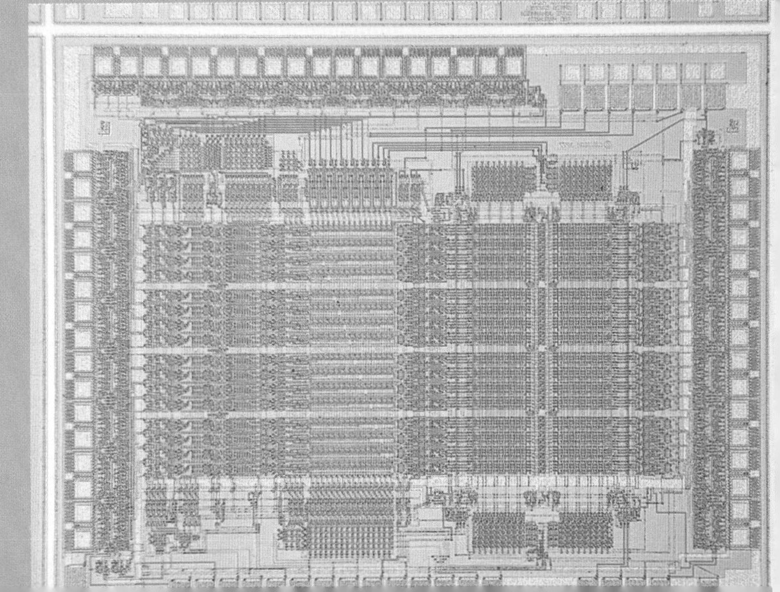

MOORE'S LAW

In 1965, *Electronics* magazine asked Gordon Moore, the research director of Fairchild Semiconductor, to write an article on the future of the integrated circuit industry. As Moore looked at the number of active components engineers had been able to put on chips since their invention by Jack Kilby and Robert Noyce, he noticed a linear progression. In 1959, there was one component; in 1964, there were 32; and in 1965, there were 64. He plotted these data on a piece of paper and, he recalled, "I just drew a line. It doubled every year for 10 years." This observation, that the number of components able to be put on a single chip doubled every year, became known as Moore's Law.

Moore's Law has been revised a few times. In 1975, he updated the rate of doubling to once every two years. Later, one of his colleagues at Intel, the company he helped found in 1968, applied the "law" to computer performance—speed—rather than strictly the number of components on a chip. (The two measures have always been linked.) Eventually the difference between the two predictions—doubling every year and every two years—was split, with people generally accepting a doubling rate of once every 18 months.

Moore's Law has proved remarkably accurate over the past 35 years. Continued innovations in manufacturing techniques have allowed components to be etched into silicon wafers in ever-increasing numbers. Using ultraviolet light, which has a smaller wavelength than visible light, chips are now made with features as small as .15 millionths of a meter. Current memory chips contain about a billion components; the latest generation of microprocessors has about 10 million.

Some computer scientists think that the era of Moore's Law may soon come to an end. The size limits of etching with light are being reached, and if the ultrathin connections between components get any smaller, the odd effects of quantum physics may impede the flow of electrons. However, research continues on alternative technologies, such as quantum processing, which depends on those quantum effects, and DNA computing, which may someday harness DNA helixes for data processing. If these or other new technologies pan out, the rate of change in computing might even get faster.

Carver Mead at bat during a playday in May, 1947. The schoolhouse in the background held about 20 students in grades 1–8.

Graduating class of 1948 at Chawanakee #1 school, Big Creek, California. Carver Mead is on the left.

Mead speaks in awe of the professors he had at CalTech in the 1950s, and he takes pride in his teaching at that institution over the past four decades.

At the time, many physicists and computer scientists were making predictions about how small a chip's transistors could be; their estimates were in the range of 10 microns. Using his own research into electron tunneling, Mead realized that "[their numbers] just didn't make any sense. They were basically invoking the principle that you can't change anything as you make the devices smaller. And of course you have to change some things. . . . When you did that, it turns out that everything got better. The devices got smaller, the power they dissipated got a lot less so the heat generated by the chip went down, and they worked a lot faster."

Mead predicted—in a talk he gave in 1968 and papers he published a few years later—that semiconductor devices could eventually be made as small as .15 microns thick. And while his colleagues disbelieved him at first, that size is just now being reached in the most advanced semiconductor fabrication plants.

This proof of how small transistors could get—and how much faster they might become—was an enormous advance in the understanding of semiconductors. But Mead's discovery raised as many questions as it answered. With transistors shrinking to .15 millionths of an inch, designers could fit tens of millions of them on a single chip. While this promised much more powerful chips, it was also a virtual impossibility, given the state of chip design at the time. To make a chip, one had to draw the circuit, then hand-trace it with a razor blade on a special silk-screening material. "It was all done by hand and enormously labor-intensive," Mead says. "And error-prone. Terribly error-prone."

Mead explains that whenever a new technology is introduced, be it the electric motor or the transistor, people first use it as a simple replacement for the old one. "So transistors were used as replacements for vacuum tubes," he says. "It wasn't until Bob Noyce figured out how to do a true integrated circuit that we got the idea, 'This is really a different medium.' The real boundary between ordinary integrated circuits and [modern ones] was when the design problem got to be something you couldn't keep in your head anymore."

In 2000, Mead's company Foveon unveiled several high-end digital cameras based on his imaging technology. The color camera's pictures are as good as 35mm photographs; a black-and-white camera produces even higher quality pictures.

He began thinking about how to automate the process. The answer was to create a silicon compiler. Like a software compiler, Mead's program, devised with the help of some of his students, let a designer describe what he or she wanted on a chip in a structured, easy-to-understand language. This language was then translated by a computer into the lines that make up a silicon circuit, and was output on a high-resolution plotter. The plotted design could then be etched onto a silicon chip.

Mead's approach, now called VLSI, or very large-scale integration, was the advance that kicked integrated circuit design into high gear. Most of the microprocessors that fueled the personal computer revolution were created with VLSI, and almost all of today's advanced chips are designed using the basic method he outlined in 1971 and expanded into book form, with coauthor Lynn Conway, in 1979.

In the 1990s, Mead turned his attention—and his frustration at people doing things the wrong way—to digital imaging. Most digital cameras begin with video technology, usually a CCD (charge-coupled device) image sensor much like those used in camcorders, rather than the principles of film. "It isn't photography," Mead says. "It's an insane way to make pictures. You've thrown away two-thirds of the light, you've thrown away two-thirds of the information, and [you have to] spend all of this computing trying to get it back when the right way to do it was to not throw it away in the first place."

In September 2000, Foveon, a company Mead founded, introduced a new generation of digital cameras that use a proprietary, silicon-chip-based image sensor that allows for detail and tonal range far greater than traditional digital cameras. Mead explains that people have known that silicon was light-sensitive for decades; in the 1960s, some engineers even proposed making photo sensors with it. However, it was only when manufacturing techniques allowed for very small transistors and very dense chips—an outgrowth of Mead's own work in the late 1960s—that silicon became practical to use in photography.

Foveon's image sensor consists of a two-inch-square CMOS chip that can capture a 16-million-pixel black-and-white image. What's more, the tonal range of the chip exceeds that of professional-quality film. Foveon also makes a color version, using a filter that directs the three primary colors of light to three separate image sensors, that boasts a resolution of 4 million pixels. Mead's new cameras, made in partnership with Hasselblad, are just now reaching professional photographers.

While there are still intriguing technical challenges for Mead at Foveon, he is also working on several "next" projects, including a silicon learning system, or chips whose programming can physically change as they learn new processes or functions. This is in part a continuation of the work he began in the 1960s, looking at the basic physical properties of silicon, but it is also an extension of the research he did in the 1970s and 1980s into creating computational analogs to the human (or animal) brain. Still, after almost twenty years, this remains a huge challenge. "People have always enormously underestimated the computing power that happens in a brain," Mead says. "That's just a big mistake. There's still a huge amount we don't know about what happens in the brain. And it's very dangerous to just assume that what we're doing is just equivalent in some way."

STEVE WOZNIAK

Steve Wozniak only wanted to impress his fellow enthusiasts in the Homebrew Computer Club. He and the others were building "microcomputers" with the new microprocessors just introduced by Intel and Motorola. He dubbed his first machine, built in 1976, the Apple. While it was impressive, it was his next creation, the Apple II, that was destined to bring computers into ordinary homes. Powerful, expandable, easy to use and program, and capable of displaying high-resolution color graphics, the Apple II opened people's eyes to what a computer could do, and why they should own one. By the time the last Apple II was produced in the early 1990s, there were millions of Wozniak's invention in homes, schools, and businesses.

Wozniak was born at the right time and in the right place for a future computer genius: in San Jose, California, in 1950. The area extending from San Jose north to Mountain View and Palo Alto was already a center for technology, home to Hewlett-Packard, the NACA (later NASA) Ames Research Center, and several aerospace companies. Over the next few decades, the academic draw of Stanford University and more and more technology companies brought thousands of engineers and scientists to the region.

In 1958, Wozniak's father, an electrical engineer, took a job at Lockheed and moved the family to nearby Sunnyvale, even closer to the core of what would become Silicon Valley. As he grew up, Wozniak fell in love with mathematics and electronics. He loved to read about Tom Swift, Jr., a teenage inventor/engineer who got involved in all sorts of adventures, often in outer space. He also subscribed to *Popular Electronics* and began tinkering with electronics. There were other technically minded kids in the neighborhood; together they made an intercom system to connect their houses.

When Wozniak was 11, he built his first computer, or something like a computer. With hundreds of components laid out on a three-by-four-foot piece of plywood, he made a machine that could play tic-tac-toe. Although his father would sometimes offer advice or help him get parts, for the most part Wozniak taught himself. In high school, Wozniak became even more involved with electronics. When he had advanced beyond the school's electronics class, he be-

When Wozniak was 11, he built his first computer, or something like a computer: with hundreds of components laid out on a three-by-four-foot piece of plywood, he made a machine that could play tic-tac-toe.

gan working for free at Sylvania, learning to run and program the company's minicomputer. He continued to design his own primitive computers, although by this point he had gone far beyond his friends. Nobody else "did this design stuff with me," he said later. "I had nobody to even show my designs to."

In 1968, Wozniak graduated from high school and began attending the University of Colorado in Boulder. He could only afford the school for a year, though, then returned to California, where he got a job at one of the many new electronics companies that were starting up. In his spare time, Wozniak continued to try to build his own computer. In 1971, he made one that worked so well that his mother called in the local newspaper for a demonstration. Unfortunately, when the reporter showed up, the computer shorted out with a puff of smoke.

Later that year, Wozniak returned to college, this time at the University of California at Berkeley. Although Berkeley was still a hotbed of radical politics, Wozniak devoted his time to studying electronics and computers. There were plenty of ways to get in trouble with electronics, however. He and his new friend Steve Jobs had heard about something called a "blue box," a device that emitted a tone at a particular frequency. When used correctly, the device would let users make free phone calls, and even listen in on other people's conversations. Always interested in jokes and pranks (even illegal ones), Wozniak set out to build his own blue box. One night, he tried to call the pope, and actually got through to his secretary. Fortunately, the pope was asleep.

Wozniak and Jobs had been introduced to each other by a mutual friend in Sunnyvale. In many ways, the two men were opposites; Wozniak was a dedicated geek, losing himself for hours or days in his electronics designs, while Jobs was a talkative visionary, eager to spread the word of technology, even if he wasn't as good as his friend at creating it. Their first business partnership came about with the blue box; Wozniak was pleased just to have made it, while Jobs wanted to sell them. They eventually built and sold about 50 of them.

The key to the personal computer is the microprocessor, which is essentially a computer etched onto a small silicon chip. The first microprocessor was invented in 1971 by Ted Hoff at Intel. By the mid-1970s, prices for the chips had begun to come down and people started to find uses for them. In January 1975, a company in Albuquerque announced the world's first personal computer: the Altair. Today, an Altair would be barely recognizable as a computer. Commands and programs were input by flipping toggle switches, and a few red LEDs were its only way of communicating back. Still, it was the first computer that was meant for one person to use and cheap enough for one person to buy. Later that year, a small group of computer enthusiasts, inspired in part by the Altair, came to the first meeting of the Homebrew Computer Club in Palo Alto. At each meeting, people would talk about their own ideas and designs for computers, and often demonstrate them.

Wozniak began to build a computer he could show off at the Homebrew club. By February 1976 it was ready, able to run programs written in the BASIC programming language and display them on a small black-and-white TV set. His fellow computer designers thought it was a great computer; more important, his friend Steve Jobs thought they could sell them. Within a few months, they had an order for 100 of the computers. To raise money, Jobs sold his Volkswagen bus and Wozniak sold his calculator. (At the time, good calculators sold for hundreds of dollars.) Apple Computer was born.

Even with that first order, Apple was just another tiny company with a computer meant for hard-core geeks. (The Apple I

The first Apple computer, designed by Steven Wozniak in 1976, came as a kit, without a case. Some users made their own enclosures out of wood and often personalized them.

TIM BERNERS-LEE AND THE WORLD WIDE WEB

When Steve Wozniak invented the Apple II, he wanted a computer that one person could use. Previously, most computers were shared by many people, who used simple ("dumb") terminals to access a minicomputer or a mainframe, which did all the work. As more and more people used personal computers, though, they began to need ways to connect with other users. The first solution was local-area networks such as Ethernet and Appletalk, which allowed several computers in an office to communicate back and forth. In the 1990s, a much larger network came into widespread use: the Internet.

The Internet was created in the late 1960s as a way to link large university and military computer networks. For the next two decades it was used by a relatively small number of people to exchange data and messages (the first e-mail). Once information had been sent through the Internet, though, there was no easy way for others to see it.

In 1980, Tim Berners-Lee, a British computer scientist working at CERN, Europe's main physics research facility, took the first step in creating an "information space" where data could be shared. Berners-Lee wanted to be able to organize the various kinds of information he used everyday—such as CERN's phone book, his notes, and his research papers. He wrote a program that let him store different kinds of documents.

Berners-Lee also had an idea about how to access all that information. Although computers are very good at finding data stored in a logical way, they are awful at recognizing connections between documents or bits of information. Humans, on the other hand, are great at making those connections, although not partic-

didn't come with a keyboard, a display, or even a case.) But Jobs saw that Wozniak's computer had great potential, and other people in Silicon Valley did, too. Mike Markkula, a business executive Jobs met through friends at Atari, joined the company and lent it $250,000. Apple needed a better product to sell, and Wozniak was already designing it.

For the next year, Wozniak worked on the Apple II, designing it almost completely by himself. Unveiled in April 1977, it was in every way a better machine than its predecessor. It was relatively fast (it ran at 1 MHz), the BASIC programming language was built in, and through some sophisticated electronic sleight-of-hand, Wozniak had given it color graphics. Wozniak's design used the minimum number of chips, which made it both cheaper and more reliable.

The Apple II was also friendlier than other computers. Sporting the company's multicolored logo and packaged in a distinctive beige plastic case, it provoked none of the anxiety then associated with computers. With the Apple II, Wozniak explained, "the phrase 'personal computer' came to mean one that a single person used, but it also meant one that was acceptable in homes, with a non-commercial appearance."

The computer was an almost instant hit. Although it initially had only 4K of memory and its programs had to be stored on cassette tape, within a couple of years it had a full 48K of RAM and an inexpensive floppy disk drive. In 1979, it gained another crucial advantage over other personal computers: Visicalc, the first spreadsheet program. Businesses found this new tool essential. The software finally answered the question, What do I do with a computer? Visicalc was a runaway best-seller; some people give it credit for much of Apple's success.

The Apple II was Wozniak's last great computer. He stayed at Apple for several years, refining succeeding generations of the Apple II and then the Apple III. In 1981, he had just begun to work with the team that would create the Macintosh computer when he was involved in a plane crash. His injuries gave him amnesia and made it difficult for him to concentrate. When he recovered, he decided he'd rather spend his time (and money—Apple had made him a multimillionaire) with his family and on other projects.

Today, Wozniak's devotes much of his time to kids and education. In the 1980s, he helped found the Tech Museum of Innovation, a child-oriented science museum in San Jose. A few years later, he discovered the joy of teaching them himself. Starting with a small group of his son's friends, he began showing kids how to use computers, and how they work. Now he leads classes at his home and at local schools, and he has donated equipment for many of the school district's computer labs. "I was born to teach," he told a biographer. "I plan to continue to be a fifth-grade teacher of computers and just get better and better at it."

ularly good at holding large amounts of data. Berners-Lee wanted to make a system that allowed for a permanent "web" of connections between information.

Over the next decade, Berners-Lee refined his ideas on how to link documents, and then found a way to make them available to many people at once. There were two parts to his solution, which made up what he called the World Wide Web. He created a standardized way to make documents readable and linkable by using a simple computer language called HTML ("hypertext markup language"). And he set up a "server"—a computer that CERN users could access that would "serve up" the document they were looking for.

about 100 "hits" a day; by 1993, that number had grown to 10,000. It wasn't quite the Web people know today; those early users' browsers could display only text. In early 1993, however, an American student named Marc Andreesen released a new browser, called Mosaic, which could display Web "pages" with different fonts, colors, and graphics.

In 1994, Andreesen cofounded Netscape and began distributing his new browser free of charge. Within a few years, millions (and then hundreds of millions) of people were on line. The combination of information, commerce, entertainment, and communication that the Web brought home has made computing an essential part of many people's daily lives.

SOURCES AND FURTHER READING

Much of the information in this book came from interviews with the subjects and from material in the archives of the Lemelson-MIT Program. The Lemelson-MIT Program's Website, The Invention Dimension (**web.mit.edu/invent**), contains short biographies of the inventors profiled here as well as many others.

What follows are other sources that were helpful in investigating the individual inventors profiled and in understanding some technical aspects of their work, as well as suggested reading for those interested in learning more about American inventors and inventing.

For detailed discussions of basic scientific and technical concepts and some biographical insights, **Brittanica.com** is an indispensable source. Marshall Brain's "How Stuff Works" (**www.howstuffworks.com**) is useful for thorough explanations of many things, notably Boolean logic and simple circuits.

The Innovative Lives section of the Website of the Jerome and Dorothy Lemelson Center for the Study of Invention and Innovation at the Smithsonian Institution (**www.si.edu/lemelson**) features illuminating essays on several of the inventors featured in this book, including Sally Fox, Ashok Gadgil, Wilson Greatbatch, and Stephanie Kwolek.

Kenneth Brown's book *Inventors at Work* (Tempus, 1988) was useful both for its interviews with several of the inventors in this book and for its incisive point of view on invention and innovation.

The United States Patent and Trademark Office has a searchable database of all patents issued since 1790, at **www.uspto.gov**.

Medicine and Healthcare

Adam, John. "Wilson Greatbatch." *IEEE Spectrum*, March 1995.

Cima, Michael, John Santini, Jr., and Robert Langer. "A Controlled Release Microchip." Press release. Massachusetts Institute of Technology, January 20, 1999.

Dale, W. Andrew, M.D. *Band of Brothers: Creators of Modern Vascular Surgery*. [Thomas Fogarty.] Appleton Communications, 1992.

Davies, Kevin, and Michael White. *Breakthrough: The Race to Find the Breast Cancer Gene*. [Mary-Claire King.] New York: John Wiley & Sons, 1996.

Kemper, Steve, "If you do not want to hear about what he does, do not ask." [Dean Kamen.] *Smithsonian*, November 1994.

Kleinfield, Sonny. *A Machine Called Indomitable*. New York: Crown, 1985.

Lloyd, Robin. "The Tissue Master Delivers." [Robert Langer.] *The World & I*, August 1999.

Maloney, Lawrence D., "Saga of a High-Tech Maverick." [Dean Kamen.] *Design News*, March 7, 1994.

Mattson, James, and Merrill Simon. *The Pioneers of NMR and Magnetic Resonance in Medicine: The Story of MRI*. Ramat Gan, Israel: Bar-Ilan University Press, 1996.
(Also at **www.heckel.org/pioneers/index.htm**.)

Regalado, Antonio. "Ideas Are Like Children." [Robert Langer.] *Technology Review*, January 1, 1999.

Straus, Eugene, M.D. *Rosalyn Yalow, Nobel Laureate: Her Life and Work in Medicine*. New York: Plenum Press, 1998.

Consumer Products

Bijker, Wiebe. *Of Bicycles, Bakelites, and Bulbs*. Cambridge: MIT Press, 1995.

Bruce, Roger, ed. *Seeing the Unseen : Dr. Harold E. Edgerton and the Wonders of Strobe Alley*. Cambridge: MIT Press, 1995.

Edgerton, Harold E. *Exploring the Art and Science of Stopping Time: A CD-ROM Based on the Life and Work of Harold E. Edgerton*. Cambridge: MIT Press, 2000.

Flatow, Ira. *They All Laughed*. New York: Harper Perennial, 1993.

Gallawa, J. Carlton. "Who Invented Microwaves?" in *The Complete Microwave Oven Service Handbook*, Upper Saddle River, N.J.: Prentice Hall, 2000.
(Also at **www.gallawa.com/microtech/history.html**.)

Kanigel, Robert. "One Man's Mousetraps." [Jacob Rabinow.] *New York Times Magazine*, May 17, 1987.

Murray, Don. "Percy Spencer and His Itch to Know." *Reader's Digest*, August 1958.

Rabinow, Jacob. *Inventing for Fun and Profit*. San Francisco: San Francisco Press, 1990.

Schatzin, Paul. "The Farnsworth Chronicles."
www.songs.com/philo/index.html.

Schwartz, Evan. "Who Really Invented Television?" *Technology Review*, September/October 2000.

Wolfe, Tom. "Land of Wizards," in *The Best American Essays of 1987*, edited by Gay Talese. [Jerome Lemelson.] New York: Ticknor & Fields, 1987.

Transportation

The Black Inventor Online Museum, "Garrett Morgan," at **www.blackinventor.com/pages/garrettmorgan.html**.

Federal Highway Administration/Department of Transportation, "Garrett Morgan: An American Inventor," at **www.fhwa.dot.gov/education/gamorgan.htm**.

Grosser, Morton. *Gossamer Odyssey: The Triumph of Human-powered Flight*. New York: Dover, 1991.

Hughes, Thomas Parke. *Elmer Sperry: Inventor and Engineer*. Baltimore: Johns Hopkins Press, 1971.

Lacey, Robert. *Ford: The Men and the Machine*. Boston: Little, Brown, 1986.

Lehman, Milton. *This High Man: The Life of Robert H. Goddard*. New York: Farrar, Straus & Giroux, 1963.

Rodengen, Jeffrey. *Evinrude, Johnson and the Legend of OMC*. Ft. Lauderdale, Fla.: Write Stuff Syndicate, 1992.

Energy and the Environment

Baldwin, J. *Bucky Works: Buckminster Fuller's Ideas for Today*. New York: John Wiley & Sons, 1996.

Djerassi, Carl. *The Pill, Pygmy Chimps, and Degas' Horse*. New York: Basic Books, 1992.

Djerassi, Carl. *Cantor's Dilemma*. New York: Penguin, 1992.

Djerassi, Carl. *Steroids Made It Possible*. Washington, D.C.: American Chemical Society, 1990.

[Ashok Gadgil.] *Physics Today*, July 1996.

Fleischer, Betsy, and V.S. Arunachalam, "Order in Disorder: Stanford Ovshinsky Talks of His Materials, Methods, and Machines." *Materials Research Society Bulletin*, November 1998.

[Ashok Gadgil.] "Water, Pure and Simple." *Discover*, July 1996.

Geary, James. "Green Machines." [John Todd.] *Time* (Latin American Edition), March 23, 1999.

Hatch, Alden. *Buckminster Fuller: At Home in the Universe*. New York: Crown Publishers, 1974.

Kremer, Gary R., ed. *George Washington Carver: In His Own Words*. Columbia: University of Missouri Press, 1987.

Neil, Kathy. "Cotton's Golden Girl." [Sally Fox.] *Hemp Times*, Summer 1997.

Todd, John, and Nancy Todd. *Bioshelters, Ocean Arks, City Farming: Ecology as the Basis of Design*. San Francisco: Sierra Club Books, 1984.

Todd, Nancy Jack, John Todd, and Jeffrey Parkin. *From Eco-Cities to Living Machines: Principles of Ecological Design*. Berkeley: North Atlantic Books, 1994.

Computing and Telecommunications

Berners-Lee, Tim. *Weaving the Web*. New York: Harper San Francisco, 1999.

Bush, Vannevar. "As We May Think." *Atlantic Monthly*, July 1945. (Also at **www.theatlantic.com/unbound/flashbks/computer/bushf.htm**.)

[Bushnell, Nolan.] There are many websites devoted to the history of video games and to Atari specifically; two of the best are Atari Gaming Headquarters (**www.atarihq.com**) and The Dot Eaters (**www.emuunlim.com/doteaters**).

[Engelbart, Douglas.] A comprehensive history of Douglas Engelbart's work on Human Augmentation, including a video of the first public demonstration of the oNLine System and the mouse, can be found at the MouseSite, **sloan.stanford.edu/MouseSite**.

[Engelbart, Douglas.] The website of Douglas Engelbart's Bootstrap Institute, **www.bootstrap.org**, explores current efforts at helping people work better, and also contains historical and biographical information about Engelbart's work.

Gemperlin, Joyce, and Tanya Scheinman. "An Interview with Nolan Bushnell," at **www.thetech.org/revolutionaries/bushnell**.

Herz, J.C. *Joystick Nation*. New York: Little, Brown, 1997.

Jett, E.K. "Phone Me by Air." [Al Gross.] *Saturday Evening Post*, July 28, 1945.

Kendall, Martha E. *Steve Wozniak*. New York: Walker and Company, 1994.

Kurzweil, Ray. *The Age of Spiritual Machines*. New York: Viking, 1999.

Leopold, George. "Beacon for the Future." [Grace Hopper.] *Datamation*, October 1, 1986.

Macdonald, Anne L. *Feminine Ingenuity: How Women Inventors Changed America*. [Erna Schneider Hoover.] New York: Ballantine, 1992.

Malone, Michael S. *Infinite Loop*. New York: Currency/Doubleday, 1999.

Mead, Carver. *Collective Electrodynamics: Quantum Foundations of Electromagnetism*. Cambridge: MIT Press, 2000.

Mink, Michael. "Inventor Al Gross." *Investor's Business Daily*, July 20, 2000.

Stanley, Autumn. *Mothers and Daughters of Invention: Notes for a Revised History of Technology*. [Grace Hopper.] New Brunswick, N.J.: Rutgers University Press, 1995.

Whitelaw, Nancy. *Grace Hopper: Programming Pioneer*. New York: W.H. Freeman and Company, 1995.

Wolfson, Jill, and John Leyba. "An Interview with Steve Wozniak," at **www.thetech.org/revolutionaries/wozniak**.

Further Reading

Basalla, George. *The Evolution of Technology*. New York: Cambridge University Press, 1988.

Brodie, James M. *Created Equal: The Lives and Ideas of Black American Innovators*. New York: William Morrow, 1993.

Brown, Kenneth A. *Inventors at Work: Interviews with Sixteen Notable American Inventors*. Redmond, Wash.: Tempus Books of Microsoft Press, 1988.

Chang, Laura, ed. *Scientists at Work: Profiles of Today's Groundbreaking Scientists from Science Times*. Introduction by Cornelia Dean and foreword by Stephen Jay Gould. New York: McGraw-Hill, 2000.

Day, Lance, and Ian McNeil, eds. *Biographical Dictionary of the History of Technology*. New York: Routledge, 1996.

Fiell, Charlotte, and Pater Fiell. *Industrial Design A–Z*. New York: Taschen, 2000.

Francis, Raymond L. *The Illustrated Almanac of Science, Technology, and Invention*. New York: Plenum Trade, 1997.

Gibbs, C.R. *Black Inventors from Africa to America*. Silver Spring, Md.: Three Dimensional Pub., 1995.

Haskins, James. *Outward Dreams: Black Inventors and Their Inventions*. New York: Walker, 1991.

Ierley, Merritt. *The Comforts of Home: The American House and the Evolution of Modern Convenience*. New York: C. Potter, 1999.

Ives, Patricia Carter. *Creativity and Inventions: The Genius of Afro-Americans and Women in the United States and Their Patents*. Arlington, Va.: Research Unlimited, 1987.

James, Portia P. *The Real McCoy; African-American Invention and Innovation 1619–1930*. Washington, D.C.: Smithsonian Institution Press, 1989.

Karnes, Frances A., and Suzanne M. Bean. *Girls and Young Women Inventing*. Minneapolis: Free Spirit Publishing, 1995.

Molella, Arthur P. "Inventing the History of Invention." *American Heritage of Invention and Technology* 4 (1988).

McKinley, Burt. *Black Inventors of America*. Portland, Ore.: National Book Company, 2000.

National Geographic Society. *Inventors and Discoverers: Changing Our World*. Washington, D.C.: National Geographic Society, 1988.

Noble, David. *America by Design: Science, Technology and the Rise of Corporate Capitalism*. New York: Knopf, 1977.

Petroski, Henry. *The Evolution of Useful Things*. New York: Vintage, 1994.

Rhodes, Richard, ed. *Visions of Technology: A Century of Vital Debate about Machines, Systems and the Human World* (The Sloan Technology Series). New York: Simon and Schuster, 1999.

Rothschild, Joan, ed. *Women, Technology and Innovation*. New York: Pergamon Press, 1982.

Science Museum of London. *Inventing the Modern World: Technology Since 1750*. New York: DK Pub., 2000.

Townes, Charles H. *How the Laser Happened: Adventures of a Scientist*. New York: Oxford University Press, 1999.

Van Dulken, Stephen. *Inventing the 20th Century—100 Inventions That Shaped the World*. London: The British Library, 2000.

Vare, Ethlie Ann, and Greg Ptacek. *Mothers of Invention: From the Bra to the Bomb, Forgotten Women and Their Unforgettable Ideas*. New York: Quill, 1989.

Vare, Ethlie Ann, and Greg Ptacek. *Women Inventors and Their Discoveries*. Minneapolis: Oliver Press, 1993.

Weber, Robert J., and David N. Perkins, eds. *Inventive Minds: Creativity in Technology*. New York, Oxford University Press, 1992.

Weber, Robert J. *Forks, Phonographs, and Hot Air Balloons: A Field Guide to Inventive Thinking*. New York: Oxford University Press, 1992.

INDEX

Profiles are indicated in color.

ACKNOWLEDGMENTS

This book is a culmination of the Lemelson-MIT Program's efforts to seek out and celebrate America's scientific and technological innovators, many of them unsung heroes of modern life. We are deeply grateful for the vision and generous support of the Lemelson Foundation and its cofounders, the late Jerome Lemelson and his wife, Dorothy, and the Massachusetts Institute of Technology. Without this unique partnership to inspire the next generation of inventors and innovators, this book simply would not have been.

There were countless contributors to this project. We must first thank Lester Thurow for his leadership in "ringing the bells" about the excitement and joy of invention, and for illustrating inventiveness as a distinctly American trait in the book's foreword. We must also thank James Burke, whose introductions to our book sections place Yankee ingenuity within the broader context of world history. We are enormously indebted to David C. Brown for telling the innovators' stories through such a lively set of biographies.

Many, many hours of research helped us conquer the daunting challenge of selecting these inventors and preparing the manuscript. Here, our greatest thanks go to Carrie Capizzi and Paul Melvin for their resourcefulness, energy and persistence. For their editorial insights and repeated readings of our manuscript, we owe a thousand thanks to Mariette DiChristina, David Lucsko, and Caryn Wesner-Early. We are also thankful for the assistance of Joshua Dorchak, Wendy Murphy, Victor McElheny, Michael McNally, Kate O'Neill, Arthur Molella, Joyce Bedi, Paul Montie, Shannon Peavey, Michael Rutter, and Lori Leibovich.

Our editors at The MIT Press, Michael Sims and Larry Cohen, showed boundless enthusiasm for the project and compassion for our mission. We could not wish for better support from a publisher. We also thank Emily Gutheinz for conveying the creative nature of invention through the book's playful design and Yasuyo Iguchi for her artistic direction.

Kristin Joyce deserves special thanks for her keen eye, superb organizational talents, and unwavering dedication to seeing through the book's production. Her command of the myriad details and deadlines was vital to this project.

We would like to thank not only the innovators in this book but the innumerable scientists, engineers, and technologists who help to shape our future through their discoveries and inventions. Without their contributions, wonders might indeed cease.

Annemarie C. Amparo
Director, The Lemelson-MIT Program

PHOTO CREDITS

Numbers preceding credits are the pages on which the images appear.

Raymond Damadian
All figures courtesy of Fonar Corp.

Thomas Fogarty
12: Paul Andrew Michaels
13 (top), 17: Fogarty Research
Remaining images courtesy of Edwards Lifesciences

Wilson Greatbatch
18, 21: Courtesy of Wilson Greatbach
19, 20: Wilson Greatbach Ltd.
22 (top): Australian Heart Association, Melbourne, Australia; (bottom) Charles O'Rear/CORBIS

Dean Kamen
24: Dean Kamen/DEKA Research
25, 26 (bottom), 29: Dean Kamen
26 (top), 28: United States Patent and Trademark Office

Mary-Claire King
30: Harley Soltes, *Seattle Times*
31: Republished with permission of Globe Newspaper Co., Inc.
32: University of Washington

Robert Langer
34: Andrew Fingland
Remaining images: Donna Coveney/MIT

Rosalyn Yalow
38, 40: From Eugene Straus, *Rosalyn Yalow: Her Life and Work in Medicine* (Cambridge, Mass.: Perseus Books, 1998), courtesy of Dr. Eugene Straus
39: Bettmann Collection/CORBIS

Leo Baekeland
All images courtesy of the Smithsonian Institution National Museum of American History

Harold "Doc" Edgerton
52, 56: MIT Museum
Remaining images courtesy of Palm Press

Philo T. Farnsworth
All images courtesy of the Farnsworth Family Archives

Stephanie Kwolek
62: Photograph by Jeff Tinsley, Smithsonian Institution
63: Michael Branscom, photographer
64: Hulton/Deutsch Collection/CORBIS
66 (top): CORBIS; (middle): Richard Cooke/CORBIS; (bottom): Ken Redding/CORBIS

Jerome Lemelson
69: Smithsonian photograph by Jeff Tinsley
Remaining photographs courtesy of the Lemelson family; patent drawings courtesy of the United States Patent and Trademark Office

Jacob Rabinow
74: Courtesy of Jack Rabinow
75: John Carnett for *Popular Science Magazine*
77: United States Patent and Trademark Office
78: Jay Velez

Perry Spencer
80 (header image): PhotoDisc
Remaining images courtesy of Raytheon Co.

Ole Evinrude
86, 89: Norwegian-American Historical Association
87: United States Patent and Trademark Office
88: Courtesy of Jeffrey L. Rodengen, author, and Write Stuff Syndicate, Inc., publisher, *Evinrude, Johnson and the Legend of OMC*. With the permission of the Boats and Outboard Engines Division, Bombardier Motor Corporation of America

Henry Ford
90: From the collections of Henry Ford Museum & Greenfield Village O.4401
91: From the collections of Henry Ford Museum & Greenfield Village, 833.158 (circa 1925)
92 (left): From the Collections of Henry Ford Museum & Greenfield Village, P.O. 6522, Jervis B. Webb Co., Farmington Hills, Mich. (circa 1925), (right): From the Collections of Henry Ford Museum & Greenfield Village, P.833.38556 (1924)

Robert Goddard
All images courtesy of Clark University Archives

Paul MacCready
All images courtesy of AeroVironment/Paul MacCready

Garrett Morgan
All images courtesy of The Western Reserve Historical Society, Cleveland, Ohio

Elmer Ambrose Sperry
115 (patent drawings): United States Patent and Trademark Office
117: Computer Museum of America
Remaining images courtesy of Hagley Museum and Library

George Washington Carver
120 (header image): PhotoDisc
120 (portrait), 121, 123: Tuskegee Institute
122: Iowa State University Library/Special Collections Department

Carl Djerassi
124 (header image): PhotoDisc
124: Photograph Courtesy of Jonas Grushkin
Remaining images courtesy of Carl Djerassi

Sally Fox
All images courtesy of Cary S. Wolinsky

Buckminster Fuller
137 (patent drawing): United States Patent and Trademark Office.
Remaining images courtesy of The Estate of Buckminster Fuller, Sebastopol, Calif.

Ashok Gadgil
140 (header image): PhotoDisc
140 (portrait): Roy Kaltschmidt, Lawrence Berkeley National Laboratory
Remaining images courtesy of Lawrence Berkeley National Laboratory

Stanford Ovshinsky
All images courtesy of Energy Conversion Devices, Inc.

John Todd
151: Courtesy San Francisco Museum of Modern Art
Remaining images courtesy of John Todd

Nolan Bushnell
156: Christopher Gardner, Metro Newspaper
157: Katherine and Isaac Esterman/CORBIS
158: Atari Historical Society
160: Uwink

Douglas Engelbart
164: Christine Alicino
Remaining images courtesy of Bootstrap Institute

Al Gross
All images courtesy of Ethel Gross

Erna Schneider Hoover
172, 174: Property of AT&T Archives. Reprinted with permission of AT&T.
173: Minnesota Historical Society/CORBIS

Grace Murray Hopper
178: Courtesy of Naval Historical Foundation
179 (top):Minnesota Historical Society/CORBIS
179 (bottom), 180: Courtesy of Naval Historical Society

Raymond Kurzweil
182: © Michael J. Lutch
Remaining images courtesy of Kurzweil Technologies

Carver Mead
188: Jon Brennis
189, 190, 191: Courtesy of Carver Mead
192: Courtesy of Foveon, Inc.

Steve Wozniak
194: Photograph by Alan Luckow
196: Smithsonian Institution